SpringerBriefs in Business

For further volumes:
http://www.springer.com/series/8860

Mary J. Cronin

Top Down Innovation

 Springer

Mary J. Cronin
Carroll School of Management
Boston College
Chestnut Hill, MA, USA

ISSN 2191-5482 ISSN 2191-5490 (electronic)
ISBN 978-3-319-03900-8 ISBN 978-3-319-03901-5 (eBook)
DOI 10.1007/978-3-319-03901-5
Springer Cham Heidelberg New York Dordrecht London

Library of Congress Control Number: 2013956652

Printed on acid-free paper

Springer is part of Springer Science+Business Media (www.springer.com)

Contents

Abstract

Top-Down Innovation analyzes the innovation leadership role of chief executives and top managers. It recommends that organization leaders apply an integrated framework to all innovation initiatives, using data as a vital connector and an essential catalyst for capturing value from innovation. To help managers align innovation efforts with organizational goals and vision, it presents three core innovation strategies: Transformative Architects, Nimble Pacesetters, and Power Practitioners. Synthesizing the lessons learned from companies that have mastered innovation over time, it provides a new perspective on the role of market-leading companies and top management in driving innovation. In contrast to theories of technology disruption and business model innovation, this study concludes that vision and leadership at the top are the decisive factors in long-term innovation success.

The case study chapters analyze innovation leadership at Amazon, Ford Motor, Netflix, Nokia, and Stanford University. Each case study characterizes the core strategies implemented by the chief executives of these organizations at critical junctures for innovation management. A concluding section summarizes the pros and cons of the practitioner, pacesetter, and architect strategies and suggests questions that managers should ask to assess their current innovation strategy.

Keywords Top-Down Innovation • Innovation leadership • Strategic innovation • Power Practitioner • Nimble Pacesetter • Transformative Architect • Amazon.com • Ford Motor • Netflix • Nokia • Stanford University • MOOCs (Massive Open Online Courses)

Chapter 1
The Challenge of Innovation

Products are getting smarter. Business models are more volatile. Customers are connected and empowered. The pace of technical change and the volume of data pouring into companies are unprecedented. Innovative winners will harness these trends to become global icons. The losers will struggle to stay in business and many of them will fail. Welcome to the world of innovation management in the twenty-first century.

If using the latest technology equaled innovation success, managers would simply choose from the endless stream of digital solutions available for implementation. Capturing the business benefits of innovation, however, is not so simple. In fact, even inventing a new technology is no guarantee of market success. As more organizations compete to launch breakthrough products, the amount that companies spend on research, development and other innovation-related activities continues to increase, but the chances of a major payback remain dauntingly low. Corporate spending on innovation reached an all time high of $603 billion during 2011. Yet nearly half of the organizations Booz & Co surveyed about their return on investment rated themselves as only marginally effective at reaping the expected rewards of innovation such as revenue growth, profitability, and turning ideas into successful new products. A Booz & Co senior partner noted that there is no long-term correlation between the amount spent on innovation and an organization's overall financial success (Jaruzelski et al. 2012).

Nonetheless, a seemingly insatiable market appetite for the latest and greatest digital products and services has fueled the growth of multi-billion dollar market leaders and an estimated $8 trillion in worldwide economic activity over the Internet. That amount pales in comparison to predicted economic benefits from the continued march of technology innovation during the next decade. "Disruptive technologies: Advances that will transform life, business, and the global economy," a 2013 report from the McKinsey Global Institute, analyzes the potential economic impact of 12 of today's early stage technologies, including the mobile internet, cloud computing and advanced robotics and estimates that these 12 innovations could account for between $14 trillion and $33 trillion in new economic activity annually by 2025 (Manyika et al. 2013).

M.J. Cronin, *Top Down Innovation*, SpringerBriefs in Business, DOI 10.1007/978-3-319-03901-5_1, © The Author 2014

With such outsize rewards at stake, it's no wonder that organizations and those who manage them are fixated on innovation. More top executives are taking a personal interest in guiding their company's innovation strategy, overseeing the implementation of specific processes for fostering innovation, or funding the development of new products and business models. Undaunted by previous project failures, managers continue to seek better strategies, new insights, and more effective means to innovate. Investors and customers expect market leaders to release breakthrough products and services at ever lower cost points across industries as varied as telecommunications, publishing, automotive, and entertainment.

Companies that fail to deliver are often penalized by the loss of customer loyalty, market share and valuation. Blackberry and Nokia, the mobile phone market leaders of the previous decade, fell behind in the global race for mobility market share when they misjudged the disruptive impact of applications designed for touch screen smart phones. While iPhone and Android devices won over consumers, Blackberry and Nokia management focused on incremental product innovations, falling into the trap that Clayton Christensen has dubbed the Innovator's Dilemma.

Christensen's research highlights an innovation paradox that plagues market leaders, especially in technology-intensive sectors. The more successful a company's products become, the harder it is for that company to adapt to radically different products and business models. Innovative product changes often don't appeal to the most profitable customers of market leaders. Legions of loyal Blackberry users, for example, loved their device keypads and didn't want to transition to typing on smart phone touch screens. The preferences of the existing customer base make many market leaders unwilling to abandon products and technologies that have propelled past growth. They try to hold onto customers by improving their existing products, but the availability of lower cost or better designed alternatives ultimately defeat these efforts. As the name of this problem implies, companies that get trapped by the Innovator's Dilemma have typically been innovation leaders at some point—but they fall behind by hanging on too long to outmoded products and technologies (Christensen 1997). Investing billions in R&D is no protection against the Innovator's Dilemma. Nokia spent $7.8 billion, amounting to 14.5 % of its sales revenue on research and development in 2011 to rank among the top ten global R&D spenders but it could not reverse a downward spiral in global market share (Jaruzelski et al. 2012).

The Innovator's Dilemma isn't the only threat to high flying market leaders. Keeping up with the inexorable rise in expectations for major new features with each product rollout can also take a toll. Apple, an acknowledged leader in innovative device design for the past decade, has lost some of its luster in the years since its record-breaking launch of the original iPhone and iPad devices. Analysts question whether it is possible to maintain the high bar that Apple has set for itself in hardware innovation and tend to punish the stock when new devices don't display significant advances. As one analyst put it, "Management knows it and so does Wall Street: the year-to-year viability of a company depends on its ability to innovate" (Nagji and Tuff 2012).

Given the rapid pace of technical advances and the hit-or-miss record of return on R&D investments, it is time for managers to reassess the role of innovation inside their organizations. In the face of such daunting challenges, where should managers look for guidance? The optimal strategy for organizational innovation has been debated for decades, along what the role that top executives should play in leading innovation efforts. This Innovation Brief provides essential insights for managers who need to understand how the escalating pace of technology innovation impacts their organization. Using case studies of the long-term successes and missteps of global innovators, it analyzes how successful innovation leaders design and carry out a fully integrated organization-wide innovation strategy.

Top-Down Innovation analyzes the innovation leadership role of the chief executive and top managers. It recommends that organization leaders apply an integrated framework to all innovation initiatives, using data as a vital connector and an essential ingredient for capturing value from innovation. To help managers align innovation efforts with organizational goals and vision, it presents three core innovation strategies: Transformative Architects, Nimble Pacesetters and Power Practitioners.

The case study chapters analyze innovation leadership at Amazon, Ford, Netflix, Nokia and Stanford University and characterize the core innovation strategies used by each of these organizations. A concluding section summarizes the pros and cons of the Practitioner, Pacesetter and Architect strategies and suggests questions that managers should ask to assess their current innovation strategy.

Core Strategies: Practitioner, Pacesetter and Architect

The three core innovation strategies exemplified by the companies analyzed in this Brief are the *Power Practitioner, the Nimble Pacesetter and the Transformative Architect*. The primary characteristics of these core strategies are illustrated in Fig. 1.1.

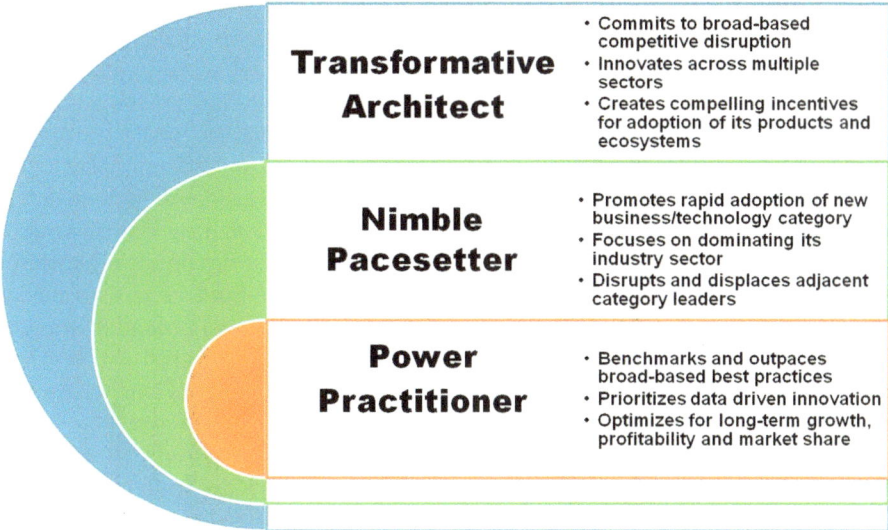

Fig. 1.1 Core innovation strategies

The Power Practitioner

Radical industry transformation, competitive disruption, and launching pioneering product categories are no guarantee of profitable business outcomes. As we have seen, high cost research and development programs often fail to deliver the expected results. In contrast, the Power Practitioner strategy is focused on long-term growth and profitability. Power practitioners do not typically aim for disruptive business models since they are well positioned to thrive within their current competitive landscape. Practitioner companies may not attract a lot of media attention, but they can reap outsize returns for their innovation investments by prioritizing innovation projects designed to grow revenues and increase market share within their industry sector.

Power practitioners benchmark themselves against the pacesetters in multiple industries. They organize internal processes to outperform the best practices in their own industry sector and adopt global best practices in the areas with the greatest impact on their company's market performance. They continually track technology, business and ecosystem innovations to ensure that they take full advantage of process improvements and technology platform advances outside of their sector. As a result of their proactive benchmarking, data-driven decision analysis, and organization-wide adoption of best practices, practitioners tend to be near or at the top of their sector in terms of productivity and profitability.

Power practitioners use data-driven decision models to identify opportunities for accelerated growth and competitive advantage. When such opportunities emerge, they are ready to invest in focused innovation efforts, for example to launch new service categories that create market differentiation or improve margins. This core strategy works well for innovation management in a broad range of organizations, from startup companies entering well established industry sectors to global enterprises seeking performance improvements.

Ford Motor used the Power Practitioner strategy to stop a downward market spiral, restore profitability and launch an innovative digital services platform, the FordSYNC. Ford fell far behind automotive industry benchmarks in the mid 2000s. By 2006, it was costing Ford thousands of dollars more than the industry average to produce each vehicle. Ford's product range included so many different vehicle platforms that its manufacturing processes were inefficient. Faced with a multi-billion dollar annual loss, Ford hired a new CEO and committed to company-wide innovation, starting with reestablishing itself as a leader in manufacturing process innovation. As the "Ford Fast Tracks Innovation" chapter discusses, Ford executives made a forward-looking decision to adopt an open platform for the FordSYNC program allowing drivers to use their smart phones to access in-car apps and services. The FordSYNC developer and platform ecosystem is modeled on Google's Android strategy, putting it in the vanguard of digital innovation for automakers.

New market entrants can excel as Power Practitioners. Wayfair, a Boston-based online retailer of home goods, has grown from a self-funded startup to a projected $900 million in 2013 sales by pursuing a power practitioner strategy from day one. The company's cofounders, Niraj Shah and Steve Conine, researched potential categories for an online retail venture. The data pointed to home goods—a $500 billion shopping category that had a huge upside potential for online growth. The founders were confident that they could outperform the competition with a combination of targeted innovation and industry best practices.

Wayfair launched as CSN Stores in 2002 and proceeded to build the largest online selection of home furniture and accessories and the industry's best performing vendor direct fulfillment network. For its first decade, CSN Stores was a collection of separately branded niche stores, each one offering extensive selection in a particular type of home goods from beds to bar stools, dining rooms to patio furniture. The company designed its own integrated vendor management, order, fulfillment and customer service platform as the unified back end for hundreds of separate stores, tracking every order, monitoring every vendor, and calculating the fastest, most cost-effective means of delivery. By analyzing performance and sales data from each niche store, the founders charted a profitable growth path to launching over 200 online stores selling over one million products by 2009, then pivoted to pull all the CSN niche stores together under a single Wayfair home goods brand.

At each stage of growth Wayfair has benchmarked itself against industry and digital pacesetters. The founders, top managers, and very Wayfair employee use the data from the company's integrated system to make daily decisions that improve performance and increase growth. When high end clothing retailers launched lucrative online private sales, Wayfair innovated to adapt this model to furniture and home goods and created the Joss & Main private sales site—a new brand that has become the fastest growing area of the company. Wayfair's Power Practitioner strategy is propelling it to surpass the billion dollar annual sales level in the next year.

The Nimble Pacesetter

Nimble Pacesetter executives harness innovation as a path to industry leadership. Pacesetter companies align with an emerging technology as early adopters, betting that they can leverage its mass market adoption to dominate a high growth business category. Their goals are rapid growth leading to market dominance in their chosen sector. As technology and category innovators, Pacesetters are intrinsically disruptive. If their technology and business model become dominant, the existing market structure will shift and incumbent leaders will often be sidelined. The Pacesetter strategy is a good match for venture-funded startup companies. Pacesetters also include well-established companies who are on the alert for innovation-based opportunities to enter adjacent markets because growth has slowed in their current industry sector,

Reed Hastings, the founder and CEO of Netflix, based his business model on a bet that millions of consumers would buy home DVD players to watch films in DVD format rather than on videotape. When Netflix launched in 1997, this was still a risky gamble. There were fewer than a million DVD players in US households that year, and the price of the player plus prices for individual film discs were prohibitively high. But Hastings' prediction that DVDs would be a consumer favorite and his decision to be a first mover in the market were spot on. Three years later there were over 12 million DVD players in US households, by 2011 over 253 million people in the US owned at least one DVD player. Netflix, as the market leader in DVD-by-mail subscription services had become a public company valued at $16 billion. To achieve this peak, Hastings had pushed his company to optimize all aspects of DVD home delivery, from a patented mailing envelope to high tech automated Netflix sorting centers that worked collaboratively with the US Post Office to streamline processing of returned DVDs and speed them to the nearest subscriber waiting for that film. Netflix also developed a sophisticated recommendation algorithm that allowed it to highlight lesser known films that matched subscriber viewing preferences and were more profitable for Netflix to provision.

Nokia, a century old Finnish manufacturer, also aligned itself with a rising star technology—the GSM mobile telephony network and compatible mobile phones. Nokia launched its first GSM mobile phone in 1992, when a standardized mobile phone network was just taking off in Europe. By 1998 there were 500 million GSM subscribers worldwide. By the mid 2000s, Nokia had become the dominant player in a market that numbered over two billion mobile subscribers.

The Pacesetter strategy has two critical risks. The first is betting on a technology or new business category that never achieves mass market acceptance, or does so long after the early ventures have failed. Most of the automotive startups that anticipated a rapid adoption of plug-in all electric vehicles during the past decade have not survived the current modest pace of consumer adoption of plug in electric cars.

The second risk stems from a very successful bet, such as the one Hastings made on DVD technology. Even the most widely adopted technologies have a limited lifespan, putting the companies that become market leaders for that technology square in the sights of the Innovator's Dilemma. It is extraordinarily hard for market leaders to shift away from a technology that is closely intertwined with customer preferences. It is even more difficult for the pacesetter in one sector to make a graceful and profitable transition to a new technology. Yet the ability to stay nimble is the key to long-term Pacesetter survival. As discussed in Chap. 4, Nokia never managed to transition to the new mobile economy of Internet-connected smart phone operating systems and applications. Reed Hastings, acutely aware of the risk of hanging on to the DVD format for too long, pushed Netflix from DVD format into the streaming media entertainment sector. It was far from graceful, as Netflix lost $12 billion in market valuation and a year's worth of subscriber growth. Chapter 3 analyzes how a nimble Netflix survived and recovered to compete in another innovation landscape.

The Transformative Architect

Industry transformation is a high risk, high reward strategy. Corporate bone yards are packed with the remains of companies that aimed for transformation and missed their target. Full scale market transformation takes time, relentless focus and extensive investment. It requires visionary executives who can articulate their long term goals to employees and inspire loyalty during inevitable reversals, delays and downturns.

Transformative Architects need deep pockets and committed stakeholders to succeed in reshaping entire industry sectors and harvesting the opportunities created by disruptive innovations. Many executives do not want to take these risks, given the extremely long odds against success. But even when companies are not aiming to be Transformative Architects, managers need to understand how these innovators will impact their industry's competitive landscape.

Transformative architects thrive on disruptive market changes—Apple, Google and Amazon are eager to launch new technologies, products and services that change the established models for doing business. An important feature of their strategy is becoming an Architect in building innovative platforms, processes and business models that will catalyze new market growth for a broad group of ecosystem partners. Apple achieved this in its coupling of the iPod with the iTunes store and even more by demonstrating the importance of an app developer ecosystem and marketplace for smartphone adoption. Microsoft's march to become the dominant operating system for personal computers and the provider of its most popular office applications had a similar impact on the business and technology ecosystems of the PC era.

As Chap. 5 describes, successful Transformative Architects incorporate the skills and data driven culture of the Power Practitioner with the focused drive of the Nimble Pacesetter to dominate in selected markets. A unique characteristic of the transformative architect is the commitment to building innovative platforms and services that provide growth and innovation opportunities to other organizations, both partners and potential competitors.

Top managers are not just responsible for selecting and articulating the most appropriate core innovation strategy for their organization. They must also lead an implementation effort that reaches all employees and areas of business. Consistent innovation requires hard work and sustained attention from chief executives. In his classic article "The Discipline of Innovation," Peter Drucker calls hard work and focus the foundation of innovation, "Above all, innovation is work rather than genius. It requires knowledge. It often requires ingenuity. And it requires focus… when all is said and done, what innovation requires is hard, focused, purposeful work" (Drucker 2002).

The Top Down Innovation Framework helps to map the core strategy into the day to day work of managing innovation across the organization.

A Data-Centered Integration Framework

Figure 1.2 illustrates the building blocks of an integrated framework in which data connects the major components of innovation across the entire organization.

Opportunities for innovation occur in every part of a company, not just in the traditional areas of products, services and business processes. Technology platforms

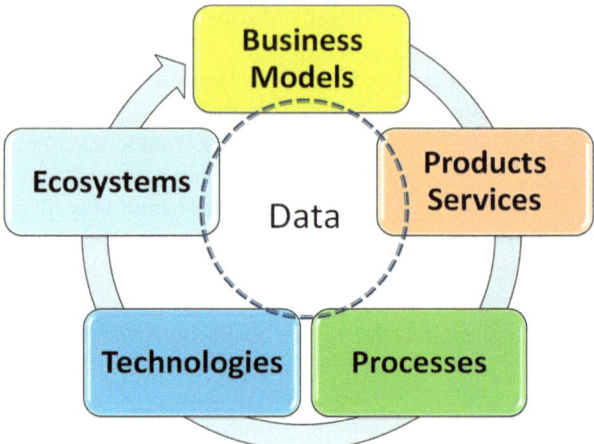

Fig. 1.2 Integrated innovation framework

based on Internet and wireless networks have opened up low cost channels for experimentation and prototyping that bypass traditional IT and management structures. Individually, managers in different business areas may come up with highly innovative ideas and practices. But stand-alone innovations that impact only one or two business areas are never optimal. Business process, new product development and business models need to be rethought as highly integrated elements of a cohesive system of innovation. It's essential to connect, analyze and utilize the data generated in and among each of the building blocks to maximize the benefits of innovation.

Essential Data Connections

Every organization is awash in data. Websites, smartphone applications, manufacturing processes, supplier and partner ecosystems, point of sale, social media sites, and other customer contact points all generate a torrent of data that is collected, stored—and largely ignored when it comes to innovation management. Data is the most omnipresent and least effectively used component of innovation.

Surprisingly few managers at the executive level put data and business intelligence at the center of their innovation strategy. In contrast, innovation leaders among Power Practitioners, Pacesetters and Architects ensure that data permeates the organizational approach to improving business processes, increasing profitability, deepening customer loyalty and evaluating the potential of new business models. These leaders are committed to using data-driven decision making as a critical component of innovation. New entrants can be more nimble than less data-savvy incumbents in responding to market trends and providing new services because of their affinity for strategic use of data. Larger companies that are weighed down with legacy systems are more likely to keep important data isolated in business line or functional silos rather than making it available to all decision-makers and employees throughout the company.

The massive amounts of data collected from every internal process, partner and customer touch point, mobile and networked infrastructure can be overwhelming to some companies. Strategic innovators are more likely to utilize open platforms and ecosystems to enhance their ability to collect and analyze massive, unstructured data. They use all relevant data to support faster development cycles increasing their ability to pivot to address new opportunities and threats.

There are signs that larger companies are starting to invest in innovative tools to manage data in ways that will boost the returns from other R&D spending. According to the 2013 Booz & Co global innovation survey, the largest R&D spenders are adopting digital prototyping, social media and data analysis tools in an attempt to catch up to the innovation strategies that come naturally to new entrants. Companies that have increased their adoption of digital tools and data analysis are 77 % more likely to report a favorable ROI from their overall R&D investments (Jaruzelski et al. 2013).

Processes and Ecosystems

Process improvement is a well-established business practice. Companies have been working on optimizing their value chains, internal operations and related business processes for hundreds of years. Henry Ford's invention of the moving assembly line at the beginning of the twentieth century revolutionized the automotive industry and propelled Ford Motor to industry leadership. The availability of large numbers of very low-cost automobiles also transformed the logistics and transportation industries and led to a nationwide highway system to accommodate the transformation in work and leisure habits for millions of new drivers. Wal-Mart and other retail giants optimized their inventory and logistics processes to create the big box discount store concept through innovations in supply chain and inventory to sustain the lower prices that attracted discount shoppers.

Business process improvements are so familiar that managers may expect only incremental ROI from additional process innovation. However when data from all the company's operations, customer interactions, partner and ecosystem models are integrated and analyzed, rethinking processes opens up major innovation opportunities. Business processes today are infused with data about every touch point inside and outside the organization, alerting managers to new product and service opportunities as well as highlighting options to reinvent the processes themselves. In turn, the reinvented process may open the gates for new ecosystem partners who will create market differentiation and competitive value. Or it may leverage new technology infrastructures such as cloud computing that reduce costs so dramatically that different business models should be considered.

By integrating its online retail customer data with backend inventory, logistics and distribution innovations, Amazon expanded its market share and boosted its revenues with innovations in free shipping, Prime Membership, and next day delivery. Free delivery and bountiful selection attracted more consumers to rely on Amazon for buying electronics, home goods, consumables and myriad other products.

Amazon's quest to optimize its internal processes by analyzing all the data these processes generate inspired the extension of its online retail store and fulfillment services to third party merchants, in exchange for sales commission and service fees. Merchant commissions and fulfillment services are now a highly profitable business model for Amazon. Wal-Mart and other big box stores are scrambling to catch up by shortening delivery times to match the customer expectations set by Amazon's process innovations.

Amazon built its ebook content and publishing ecosystem for the Kindle by bringing traditional publishers and individual, self-published authors together to offer a massive collection of content at all price levels. A content ecosystem with millions of free titles attracts customers who actively contribute their ratings and comments to the ecosystem, increasing customer loyalty and making the Kindle marketplace all the more attractive to publishers and authors.

Companies that do not empower their partners to energize their ecosystems as part of an integrated innovation strategy will be competing with new entrants who have decided that partner and customer collaborations are the fastest and most cost-effective way to bring potentially disruptive and transformational products and services to market.

Business Models, Products and Services

Business model innovation is attractive to top-level managers because business models seem to be most directly linked to top line revenue increases and opportunities for market leadership. However, just as the value of process innovation comes from its interconnections with the other building blocks, companies should beware of prioritizing business model, product or services innovations that are not informed by data. The most innovative business models stem from integrating an organization's processes, technology platforms and ecosystems. Transformational integration with mobile, social and online platforms allows companies to bring more differentiated services and products to market with more compelling value propositions.

Ford Motor, for example, adopted digital best practices such as opening its embedded platform to application developers for its connected car business model and services. Stanford University supported the interest of its faculty in experimenting with massive, open, online courses (MOOCs), leveraging cloud and social media platforms that brought hundreds of thousands of students together into a collaborative learning environment. The results of these experimental massive courses catalyzed the launch of Coursera and Udacity, two companies that are developing

MOOC platforms, services and disruptive business models in the higher education sector. In contrast, Nokia and Blackberry were so focused on their primary business model of making and selling mobile devices that they fell behind in developing innovative mobile services and applications. As a result, both companies were too late in attempting to create the application developer and app market ecosystem that have become essential elements of mobile innovation and consumer adoption.

Technology and Platforms

Internet, mobile, social and cloud computing platforms are not just extensions of earlier technology capabilities. As integral components of the Top Down framework, they can become launching pads for disruptive products, services and business models. Ubiquitous global connectivity via Internet and wireless infrastructure has unleashed a torrent of digital products and services over the past decade. These new and increasingly available platform innovation options have changed the rules for business innovation in a significant way that is still not factored into the innovation strategies of most of today's companies. New tools for innovation leverage cloud computing, mobile and social media expand opportunities outside of the traditional organizational structure to enable customers, partners and even casual online users to contribute to shaping new products and services.

Cloud computing and open source software provide low-cost and easily scalable opportunities to pilot and develop prototypes that can be rapidly deployed to global markets at low cost or free price points. Open innovation which leverages the availability of existing open source technologies and platforms provides a disruptive path for new entrants to compete with well established companies by launching new products and services at a rapid pace, testing market response, and forgoing revenue until a critical mass of consumers has adopted. Low-cost availability of connectivity, computing cycles and data storage through cloud services have leveled the playing field for launching potentially disruptive mass market services. Digital platforms and open source infrastructure have eroded many of the size and scale advantages that larger enterprises relied on, facilitating a massive influx of small competitors who can scale up quickly. As a result, every industry is open to disruptive new entrants.

The following chapters analyze the strategies of five organizations and their chief executives at critical junctures for innovation management.

References

Christensen, C.: The Innovators Dilemma, Harvard Business Review Press, Boston (1997)
Drucker, P.F.: The discipline of innovation. Harvard Business Review 80(6), 95–103 (2002)
Jaruzelski, B., Loehr, J., Holman, R.: Global innovation 1000: Making ideas work. Strategy + Business (69) (2012)

Jaruzelski, B., Loehr, J., Holman, R.: Global innovation 1000, navigating the digital future. Strategy + Business (73) (2013)

Manyika, J., Chui, M., Bughin, J., Bughin, J., Dobbs, R., Bisson, P., Marrs, A.: Disruptive technologies: Advances that will transform life, business, and the global economy. McKinsey (2013). URL: www.mckinsey.com/insights/business_techology/disruptive_technologies

Nagji, B., Tuff, G.: Managing your innovation portfolio. Harvard Business Review (2012). URL: hbr.org/2012/05/managing-your-innovation-portfolio

Chapter 2
Ford Finds Its Connection

> Today, we are moving from a culture that discourages innovation back to a company that celebrates it…. This company was founded by an inventor; we want to make sure that today the company is overflowing with innovators. We're going to find them, encourage them, and then we're going to reward them.
>   Bill Ford, Executive Chairman, Ford Motor Company (Ford 2006)

Ford Motor had a problem. Workers at its state of the art assembly plant were taking too long to produce each vehicle. Top management came up with a radical innovation by rethinking the standard automotive manufacturing process. Productivity soared, with the average time to produce each vehicle speeding up from over 12 h to just 90 min. With further process improvements, Ford accelerated the pace of global production to under a minute per vehicle. This was an unprecedented level of efficiency, a change that allowed the company to reduce car prices dramatically. Lower prices meant that millions of Americans could buy their first automobile. The advantages of Ford's manufacturing innovations were so enormous that they revolutionized the entire automotive industry. The process efficiencies were quickly copied by other manufacturing sectors generating similar productivity advances. The radical improvement that Ford pioneered back in 1913 was the moving assembly line, an invention that established Ford's early reputation as an architect of transformation.

Under the relentlessly inventive leadership of founder Henry Ford, the company went on to pioneer dozens of innovations in automotive production and design. By the following decade, the Ford plant could assemble a vehicle (the famous Model T) in about half a minute. Ford became the global auto industry leader, scaling up its output, further reducing its cost per vehicle, and boosting profitability. In 1927 Ford Motor produced and sold more than 15 million automobiles, representing half of the total global automotive market. Ford survived subsequent economic downturns and world wars to maintain its position as a leading global brand and one of the largest family-controlled companies in the world.

Despite this enviable century of growth, Ford Motor faced a crisis in 2006 that raised questions about its long-term corporate survival. Ford's share of the North American car market had eroded from 23.7 % in 2000 to just 15.5 % by mid decade.

M.J. Cronin, *Top Down Innovation*, SpringerBriefs in Business,
DOI 10.1007/978-3-319-03901-5_2, © The Author 2014

The company's model selection was heavily weighted toward trucks, large SUVs and sedans, increasingly out of tune with consumer preference for the fuel economy offered by smaller cars and for more economical crossover sports utility vehicles. Ironically, the company that had pioneered manufacturing efficiencies now lagged behind industry best practices. Ford's average cost to produce a vehicle was $2,500 higher than the global industry average.

In the second quarter of 2006, Ford Motor reported the largest quarterly loss in its history, $1.27 billion. This was a bitter pill for investors, as was the forecast that Ford's current performance would result in a full year loss of over $8 billion. The company's corporate bonds were downgraded to junk status and Ford had to put most of its corporate assets on the line as collateral to borrow the billions of dollars it needed to meet operating costs. Thanks to aggressive cost-cutting and a surprisingly strong Q3, the company's fiscal 2006 results were not as dire as anticipated. Nevertheless, Ford ended the year with a total loss of $2.7 billion. Annual profits, moreover, were not expected to return until 2009. And this grim outlook was about to get considerably worse as Ford sales sank along with those of the entire automotive sector in the global financial crisis of 2008.

It took 2 years of budget cuts, over 30,000 layoffs, 14 plant closings, elimination of unprofitable vehicles lines, and internal reorganizations to streamline Ford's production processes enough to stem the company's losses. Of the big three US automakers, Ford was the only one able to survive the 2008 downturn without government assistance. By 2009, Ford had achieved its short term financial goals, ending the fiscal year with a profit of $2.7 billion, the company's first annual profit in 4 years. Against long odds, Ford Motor had positioned itself for a second century of growth.

Ford has come back strong from its 2006 crisis, but the company is still a long way from matching the transformational inventions of its early days. Achieving Bill Ford's vision for a resurgence of innovation remains a work in progress. This case will analyze the strategies that Ford Motor has adopted since 2007 to restart its innovation engine and become a high growth, profitable Power Practitioner. By adopting innovation best practices from technology and ecommerce companies such as Google, Apple and Netflix, Ford is positioning itself to become a digital pacesetter within the automotive industry.

Automotive Industry Context: Incremental Innovation

The automotive industry has a long-standing track record of investing billions in research and development. Each major car maker funds an in-house research team and multiple development groups. In 2013 the automotive industry as a whole will invest over $100 billion in R&D. Volkswagen, which is the biggest spender among the 1,000 global companies surveyed by Booz & Co, will spend $11.4 billion. Toyota will spend $9.8 billion to join Volkswagen on the list of top ten top global spenders. Honda and Daimler each spend over $6 billion, making it into the ranks of the top 20. But automakers are not prominent in the list of the world's most innovative companies as ranked by the Booz survey, reinforcing the conclusion that R&D investment does not reliably translate into breakthrough innovations. The highest ranked global innovators are Apple at number one and Google at number two, followed by Samsung and Amazon. The only automaker in the top ten is Tesla Motors, the electric car maker which spent a relatively puny $300 million on R&D last year (Jaruzelski et al. 2013).

Automakers pour billions into research, but the auto industry as a whole produces incremental rather than transformational change. Year after year, leading car brands introduce new automobile models with updated navigation, safety and driver assistance features that reflect an impressive level of advanced technology. Despite all these features, the baseline characteristics of mass-market automobiles have remained stable over time and the industry itself is highly resistant to disruptive innovation. When the year's new car models are compared to each other they end up being very similar to other brands in the same price category and not very different from last year's models.

From decade to decade, carmakers put their efforts into improving the performance and safety of their vehicles through long-term engineering projects that are slowly introduced to the market. Each vehicle has multiple interconnected computers, electronic sensors and internal networks. A mid level car today typically contains hundreds of sophisticated microprocessors that manage everything from vehicle steering and braking to door locks, turn signals and seat belt sensors. High end vehicles contain an array of advanced navigational, entertainment, and driver monitoring features that support radar enhanced collision avoidance with automated steering, braking, and other advanced safety capabilities. It takes years for new features to make it from the research lab to the auto showroom. In addition to meeting safety, regulatory and performance testing hurdles, any innovations that impact core auto components will require changes in the manufacturing process and possibly an expensive adaption of existing assembly lines. Adding new vehicle components may also require supply chain adjustments. The time and expense involved in retooling for fundamental changes is a strong industry-wide deterrent to radical innovation (Cronin 2010).

Car makers anticipate little threat from disruptive new market entrants, and with good reason. The complexities of new vehicle design and manufacture, combined with the enormous capital costs of building new factories, are daunting and effective barriers to automotive start ups. Despite the hundreds of millions in venture and government funding invested in new electric vehicle makers during the past decade, few survived long enough to bring commercial models to market. Tesla, one of the innovative and potentially disruptive survivors, still faces a long uphill battle to become a well-established brand.

In this context, Ford had to develop its Power Practitioner innovation strategy from the ground up, seeking models outside the automotive industry for best practices in innovative digital services, open platforms and developer ecosystems.

Rebirth of Innovation at Ford

As Bill Ford acknowledged in his 2006 business review, Ford Motor was facing a struggle for survival. With the future of the company at stake, Ford turned first to the areas in which it had fallen behind. One of these was internal production processes. A lack of coordination in manufacturing systems and a plethora of different vehicle bodies and incompatible parts had pushed Ford's production costs far above the industry average. To reduce its cost per vehicle and bring its production processes into line with industry best practices, Ford standardized its production infrastructure worldwide. It modernized and retooled its manufacturing plants to increase process efficiencies.

Responding to consumer preferences for smaller and more fuel efficient vehicles, Ford developed a plan to roll out new or upgraded versions of 70 % of all Ford vehicles along with a new emphasis on developing hybrid and electric vehicles to serve the growing market niche of environmentally conscious car buyers. This model upgrade was accompanied by an internal program to streamline Ford's vehicle design process, eliminating older models and those that had lagged in sales. Ford selected a small number of base vehicle platforms upon which it could manufacture multiple car models. Instead of designing different cars for each of its global markets, the company used these standard base platforms with selected design and option variations to accommodate regional preferences.

Implementing such changes was essential to the success of Ford's campaign to rein in costs and regain competitive parity with leading manufacturers such as Toyota in process efficiency and cost per vehicle. Management pushed through innovations in each area of the company. Alan Mulally, the CEO whom Ford hired to spearhead the company's turnaround in 2006, insisted that managers use shared data to integrate across business areas. According to the company's data and analytics managers, Mulally's arrival created a data-driven culture, pushing the company to abandon its legacy of separate data silos to create an integrated data platform. While Ford Motor caught up with industry best practices in manufacturing and design, the company was also preparing to embark on more far-reaching innovations in automotive digital services (Hiner 2012a, b).

The rebirth of innovation that Bill Ford had promised came in a series of innovations based on best practices from the digital sector. Ford was determined to move beyond the traditional, hardware-focused, proprietary solutions typical of the automotive industry to innovate with digital applications, services, and open platforms. Top management wanted to use digital innovation to differentiate the company in the eyes of consumers. Ford's first step was its launch of the Ford SYNC platform in 2007.

Ford's embrace of an open platform and Bring Your Own Device (BYOD) strategy for connecting smartphones and their apps to Ford vehicles was a bold decision for the industry. Instead of insisting on end-to-end control of all the hardware and software in its vehicles, Ford was inviting partners to create strategic in-car services. The SYNC strategy reversed the long term trend for automakers to embed proprietary communication and navigation solutions in their high end vehicles as exemplified by GM's OnStar. A comparison of the Ford SYNC with the OnStar service illustrates the disruptive potential behind Ford's decision.

Like OnStar, SYNC is designed as an embedded platform for providing frequently used in-car features and services such as maps, navigation and turn by turn directions, voice enabled calling through a mobile phone interface, music, news and other online content and applications. But the innovation strategy behind the launch of SYNC is completely different from that of OnStar.

Rather than developing its own system from scratch, Ford partnered with Microsoft to license the Windows Embedded Automotive platform and customize it with Ford-specific options. That itself was a radical step for the automotive industry in 2007. But Ford went further. Even though the core OS and embedded software was based on Windows, Ford insisted that the new SYNC module had to be designed to work equally well with all mobile operating systems and cell phones. Consumers would use Bluetooth wireless or USB to link their mobile phones and music players to the Ford SYNC platform to access apps and entertainment. If drivers bought a new phone, or added apps, they could continue to use Ford SYNC with their upgraded device.

As is typical of new technology, especially technology that attempts to interface and interoperate with multiple hardware and software systems, the performance of Ford SYNC was far from perfect at the outset. Bluetooth wireless connections were not easy for many mobile phone owners to manage and the platform software was buggy. But compared to OnStar and other expensive built in navigation systems, the price was right. Since Ford was relying on the driver's phone to provide connectivity rather than installing an expensive embedded cellular module and antenna, it was affordable to offer SYNC as a free service for up to 3 years in selected new vehicles.

The timing of SYNC's market entrance was also right on target with the widespread adoption of a new generation of smartphones and apps. 2007 marked the debut of the iPhone and Android phones with advanced voice calling features, integrated music and Google maps navigation that were perfect for connecting to the SYNC platform. The more consumers bought advanced smartphones and loaded them with apps, the stronger the value proposition of a BYOD platform strategy for Ford. Drivers could update their phones as often as they wished, creating their own personalized mix of in-car entertainment and services to complement the Ford-provided services such as 911 Assist for automatic calls to roadside assistance in case of an accident.

In contrast, when GM developed its OnStar service in 1996 it did not even consider the open platform, BYOD option—partly because mobile phones at the time were limited to voice calling, but primarily because GM executives wanted the system's design, hardware, and user interface to be totally controlled by GM. As an innovative service for its time, OnStar was expected to build customer loyalty as well as to generate revenues through a fee for installing the OnStar option on new GM vehicles plus charging drivers an annual subscription fee for the service.

The components required for OnStar, including an embedded wireless communication module, GPS and an antenna for satellite connectivity, added considerable cost to each vehicle in which it was installed. Moreover, OnStar services were voice-enabled and relied on live operators being available 24 h a day, 7 days a week to provide turn by turn directions and safety assistance such as summoning roadside help to the scene of a breakdown. According to Chester Huber who joined GM to help design and launch OnStar and spent 14 years as its first president and CEO, GM spent over $1 billion investing in and subsidizing the OnStar service before it established enough of a customer base to become self-sustaining (Nobel 2013).

In addition to costing GM over a billion dollars, the proprietary design of OnStar required a constant and costly struggle to keep up with rapid improvements in competitive solutions. The first disruptive competitors were standalone personal navigational devices (PNDs) that became market favorites. Garmin, Magellan, Tom-Tom and other PND manufacturers offered superior mapping and navigational services featuring full color display screens, interactive touch graphics and other features that OnStar lacked. Over time, PND price points fell to under a 100 dollars for basic models, making them affordable for most drivers. The global adoption of the iPhone and Android smartphones in turn disrupted the PND market. With so many low cost and free options available, drivers saw even less value in a high priced embedded option with annual subscription fees and GM's opportunity to expand the traditional OnStar customer base declined.

Ford's decision to implement an embedded platform designed by Microsoft and the innovation of letting drivers to use their own mobile phone to provide connectivity avoided many of OnStar's problems. Partnering with other vendors and relying on a smartphone interface instead developing everything in house lowered the cost structure for delivering and updating in-car services. It also helped to extend the lifespan of the SYNC platform—an important consideration given the trend toward consumers keeping their cars for longer periods of time.

Since Ford SYNC provided standardized interfaces for all types of mobile devices, drivers can decide for themselves which apps and features are most important, updating their phones without needing a parallel update of the SYNC system. This solves the challenge of anticipating consumer tastes and smartphone capabilities at the time of car manufacture—an impossible task when device features and apps are changing so quickly from year to year. At the same time, the embedded SYNC module offers a secure connection from the phone to the vehicular network. This connection allows Ford to enable voice control of the car radio and other in-vehicle services. SYNC can also connect the phone display to a variety of safety features and vehicle performance and diagnostic information that can be used in developing third party applications.

Because Ford was not required to embed an expensive communications module or take responsibility for the satellite or cellular network conductivity, it was cost-effective to build the SYNC module into a broader group of vehicles. These lower costs also gave Ford greater pricing flexibility when it marketed SYNC and its MyTouch successor to car buyers. As of 2013, Ford SYNC is active on over five million vehicles. This compares very favorably to the estimated 5.1 million active OnStar subscribers in 2010, a full 14 years after GM's launch of its service.

Despite its strong points, Ford SYNC and especially Ford MyTouch system illustrates many of the risks faced by manufacturing companies innovating in software and services. Like much newly released software, MyTouch came to market with unresolved bugs and a propensity to crash in some configurations. So many MyTouch customers reported connectivity and interface problems that complaints over social media reached a fever pitch and Ford's overall vehicle consumer satisfaction rating suffered. Ford responded with upgrades, interactive support, and new software releases. The bottom line is that the innovations in SYNC and MyTouch are meeting Ford's goals of differentiating itself as a digitally savvy car maker and attracting new buyers. In a recent consumer survey, Ford noted that 80 % of 2013 buyers indicated their interest in a Ford vehicle was related to its reputation for advanced digital services and their desire to have a system like SYNC in a new car.

The Connected Car

At the beginning of 2013 Ford took the next step toward a fully open strategy for the connected car. It announced the formation of an OpenXC partnership that would make the Ford SYNC AppLink interface free and open to third party application developers. OpenXC lets registered users download a software development kit and the details of sensor specifications for Ford automobiles in order to encourage application developers to engage with the technology and eventually develop apps for the Ford platform.

Going even further, Ford announced that the entire SYNC AppLink platform was open and available for use by its competitors in the automotive industry. One analyst characterized this announcement as the beginning of Android for automotive apps, noting that Ford was breaking new ground by giving competitors cost-free access to the foundations of the SYNC AppLink system with no restrictions on how the system could be used in other manufacturer's vehicles. Ford's open platform offer could dramatically increase the availability and the strategic importance of applications for the automotive industry. Automakers that decided to adopt AppLink would have the advantage of joining a platform that was already well-regarded by embedded system developers, increasing the likelihood attracting innovative new ideas and app options to their ecosystems (Lavrinc 2013).

Developers responded positively to the idea of experimenting with the free AppLink resources. Since Ford's announcement, thousands have registered to download the AppLink specs. A significant number of developers have approached Ford with ideas for new services. Given the long-standing inclination toward closed ecosystems and proprietary platforms in the automotive industry, it is not surprising that the other auto manufacturers are less enthusiastic about Ford's offer. Competitive automakers expressed reluctance to drop their own proprietary systems in order to join in with Ford's ecosystem. A General Motors executive remarked that from his company's perspective adopting Ford's AppLink standard would be immensely risky. Why? The open nature of the platform could result in a loss of control by GM.

This attitude underscores the disruptive nature of Ford's open source initiative along with the potential for Ford to become an industry pacesetter if it succeeds at attracting enough partners and app developers to create the vehicle equivalent of an App Store.

The AppLink open source announcement kicked off Ford's bid to become the open-source automotive innovator, with the Ford platform as an industry-wide standard for in car services and applications. It's another example of the Power Practitioner strategy of adapting best practices from other industries. Media comparison to Android is premature, however, even though Ford's current strategy with SYNC and AppLink clearly draws inspiration from Google's open source model. To come anywhere near Android's level of success, Ford will have to redouble its efforts to attract a vibrant community of app developers and a strong following among consumers.

To generate market interest, Ford sponsored a developer competition in 2013 with prizes for the best applications and an opportunity for the general public to suggest a theme for the first round of competition. Consumer votes favored apps that would help drivers to increase fuel efficiency—an area which allowed Ford to showcase AppLink's ability to let developers tap into the data generated by various vehicle information systems.

Contest results were announced in fall 2013. Twenty-three applications designed to increase fuel efficiency had been submitted. A start up called Fuelytics won the grand prize with an app that compares the car's fuel results with similar models and Eco-Dash came in second with a graphic interface to help drivers visualize their fuel consumption in real-time. Despite the relatively low number of entrants, the contest outcome was positive for Ford. This was a contest that required developers to get familiar with the AppLink tools and to spend time deeply engaging with Ford's technical infrastructure and APIs. It extended the knowledge base needed for future development work.

Looking beyond its return to profitability based on industry best practices in manufacturing, Ford has crafted a well-integrated strategy aimed at becoming the leader of an ecosystem for in-car services and apps. It is ahead of the industry in embracing open standards and attracting software partners. Its Ford SYNC services have proven themselves to be attractive to new car buyers. As the next section will discuss, Ford has also demonstrated its skill as a Power Practitioner in leveraging social media to create a direct connection to consumers, instead of relying primarily on broadcast advertising and dealership sales channels.

Innovative Channels for Reaching Social and Mobile Car Buyers

The car dealership sales model dates back to the start of the automotive mass market. Independently owned US dealerships were established as a way for automakers to expand their reach into all geographic areas. Car dealerships provided showrooms and sales staff to encourage buyers to view and test drive new models. In addition to providing information and driving experiences, the dealership staff gave distant auto manufacturers a trusted, local face and personality. Local car dealers were typically pillars of their community who were actively involved in all sorts of civic, business and neighborhood activities. This made them excellent ambassadors for promoting an automotive brand in the early days of the industry.

The early car dealers, however, were often forced to buy inventory from the big automakers well in advance of their hope of making a sale. This cushioned the automakers from economic ups and downs and guaranteed them steady flow of sales but it left the dealers to handle excess inventory. In exchange for their services, dealers lobbied for legal protection and gained exclusive right to sell cars directly to consumers.

In the days before ecommerce, giving dealerships an exclusive channel to consumer sales made sense to automakers. It protected dealership investments, motivating them to expand their physical locations and local ad budgets. But now that ecommerce has become a multi-billion-dollar sales channel for all types of goods and services, the automakers are chafing at the dealer franchise restrictions that prevent them from selling cars directly to consumers over the Internet.

Most automakers still rely on traditional broadcast and print advertising supplemented by websites that allow consumers to explore vehicle features and configure models with a variety of different options. Online configuration lets automakers see how their potential customers think about which features are most important and how they respond to the bottom line price for a customized vehicle. But just at the point of decision, prospective customers are referred to nearby dealerships to switch from online interaction to visiting a showroom. Consumers who land on automaker websites are already interested in a particular brand. There is a large gap between those serious car buyers and the general consumer who may or may not pay attention to a 60 s television commercial.

Ford was an early adopter of social media marketing to bridge that consumer gap, using innovative social media campaigns to convey the experience of being a Ford owner to the general consumer. One of Ford's first social media campaigns served to launch a redesigned Ford Fiesta brand in 2009. To attract a younger audience to the new Fiesta, Ford launched a contest for consumers, asking entrants to describe why they would enjoy driving a new Fiesta free for 1 year. From thousands of entries, it selected 100 drivers to experience the Fiesta—and to share that experience with the world through regular blogging, tweeting, Facebook posts and videos. Winners were encouraged to talk candidly about the positives and the negatives of their experience.

The outpouring of opinions and spontaneous observations from Ford's 100 social drivers articulated the day to day experience of owning a Fiesta in ways that professionally produced commercial messages could never capture. Ford edited clips from some of the winners to put together a more traditional marketing campaign, retaining the authentic voice and videos from the participants. This approach to marketing leveraged the fact that consumers today are much more likely to trust their peers and to make purchasing decisions based on advice from friends. Positive comments on social media and candid consumer reviews of products have more influence than advertising. The Fiesta campaign presented Ford in an authentic and engaging light much more effectively than a series of ads that would be easily tuned out by young car buyers. By the end of the Fiesta campaign Ford had attracted almost 2,000,000 fans on Facebook and over 200,000 Twitter followers.

Buoyed by this successful experiment, Ford has continued to innovate in its use of social media. It decided to forgo traditional advertising and trade show channels in 2010, instead launching its new Ford Explorer on Facebook. The Facebook launch was coordinated with social and digital marketing efforts to encourage consumer interaction. Over a 4 week period, the Explorer campaign reached over 100 million people via social media, a resounding success in terms of consumer engagement at a price point that was significantly lower than the cost of a single Super Bowl ad campaign (Laskowski 2013).

Ford's early use of social media helped it to build a reputation as a fun and engaging brand in the minds of younger consumers. Ford's innovations with the SYNC platform and its support for BYOD appeal to the same generation of young car buyers and owners who are the heaviest users of social media. To complement Ford's integrated innovation strategy, social media marketing provides a data-rich view of consumer trends and preferences that Ford can integrate with the data it collects from other sources, including vehicle performance data and insights into driver habits and their use of in-car services. The next section discusses how Ford is working to analyze and apply the data generated by its open platform, application ecosystem and social media to further its innovation strategy.

Big Data Opportunities: The Car as a Service

According to Gartner Research, "to generate future profits, automakers need to make a transformational shift from manufacturing and selling physical vehicles to designing smart mobility services and analyzing the data such services generate. This shift will depend on smart, connected vehicles that generate unprecedented amounts of data" (Koslowski 2012).

Bill Ford has been promoting a similar vision of smart mobility, inside his own company and more broadly in speaking with stakeholders in the wireless and technology industries. In a 2012 keynote speech at the Mobile World Congress, Ford urged telecommunications and other technology companies to work together with automakers to connect the billion-plus unconnected vehicles on the road today. To address global transportation needs, Bill Ford envisions creating an interconnected transportation network in which vehicle-to-vehicle communication could route drivers around traffic gridlock and alert them to dangerous conditions ahead.

If Bill Ford is the acknowledged visionary leader of Ford Motor then CEO Alan Mulally is its leading Power Practitioner. Mulally engineered Ford's turnaround by realigning the company toward best-practice based innovation in every area of business. He is also the internal evangelist for integrating and leveraging data as the foundation of future innovation. Ford is currently working on big data projects to integrate the unstructured consumer opinion data generated by its Internet and social media activities with the unending stream of data that flows from the Ford SYNC Platform. Future plans include sharing that data with ecosystem partners and allowing app developers to leverage it in mobile consumer apps (Hiner 2012a, b).

Just how massive is the data collected by each vehicle in Ford's installed base of vehicles? A recent article reported that the 2013 Ford Fusion Energi plug-in hybrid has more than 70 embedded computers that process data from the dozens of sensors that are constantly monitoring the vehicle performance and every action of its driver. All those sensors produce more than 25 GB of data every hour that the car is on the road. Ford is not yet analyzing all that data to spot meaningful trends across all such vehicles, and it may be years away from having a seamless integration and analytics engine that can pull together predictive forecasts and holistic market insights based on multiple data resources. Thanks to the vision of top executives, the company has a head start working on that goal. Ford is making big data a priority in its quest to create new business models and innovation breakthroughs (McCue 2013).

Conclusion

Bill Ford is not content with being an automotive industry Power Practitioner. He has set his sights on a smart mobility future in which making and selling vehicles will be only one part of Ford Motor's business model. As he noted in a 2012 interview, "I suspect we'll always be making cars and trucks, but we may be doing something else very different as well. If we think of ourselves as a mobility company rather than just an automobile provider that really opens lots of different possibilities" (Hiner 2012a, b).

As we have seen, Ford emerged from its mid 2000s struggle for survival as an innovative adopter of best practices in the digital as well as automotive sectors. It demonstrated a commitment to innovate across the entire company and has delivered on that commitment with profitability, global growth and an open platform for the connected car and third party mobile applications. Ford is preparing to thrive in a future where connected cars become the primary business models of car makers and their partners. In this long-term future, car makers will have to offer smarter services to generate revenue and only the most innovative will survive.

Whether Bill Ford's radical vision of the automotive future ever materializes, Ford Motor has succeeded in reasserting itself as an innovative force with the resources and the vision to implement disruptive innovations in specific areas such as Ford SYNC and its open application infrastructure.

References

Cronin, M.: Smart Products, Smarter Services. Cambridge University Press, Cambridge (2010)
Ford, W.: Ford Motor Company Business Review, Ford On-Line (January 23, 2006). URL: www.drivingthenation.com/?p=552

Hiner, J.: Ford is now a 'personal mobility' company: How the comeback kids are riding tech to a new destiny. CNet (July 14, 2012a). URL: news.cnet.com/8301-11386_3-57472376-76/ ford-is-now-a-personal-mobility-company-how-the-comeback-kids-are-riding-tech-to-a-new-destiny/

Hiner, J.: Ford's big data chief sees massive possibilities, but the tools need work. ZDNet (July 5, 2012b)

Jaruzelski, B., Loehr, J., Holman, R.: Global innovation 1000, navigating the digital future. Strategy + Business (2013)

Koslowski, T.: How technology is ending the automotive industry's century-old business model. Gartner Research (September 25, 2012)

Laskowski, A.: How Ford became a leader in social media. BU Today (April 19, 2013). URL: www.bu.edu/today/2013/how-ford-became-a-leader-in-social-media/

Lavrinc, D.: Exclusive: Ford wants create the Android of automotive apps. Wired (January 7, 2013). URL: www.wired.com/autopia/2013/01/ces-2013-ford-applink-for-all

McCue, T.J.: 108 MPG with Ford Fusion Energi plus 25 gigabytes of data. Forbes (January 1, 2013). URL: www.forbes.com/sites/tjmccue/2013/01/01/108-mpg-with-ford-fusion-energi-plus-25-gigabytes-of-data/

Nobel, C.: Lessons from running GM's OnStar. Working Knowledge, Harvard Business School (March 4, 2013). URL: hbswk.hbs.edu/item/7185.html

Chapter 3
Netflix Switches Channels

> For the past five years, my greatest fear at Netflix has been that we wouldn't make the leap from success in DVDs to success in streaming. Most companies that are great at something – like AOL dialup or Borders bookstores – do not become great at new things people want (streaming for us) because they are afraid to hurt their initial business. Eventually these companies realize their error of not focusing enough on the new thing, and then the company fights desperately and hopelessly to recover. Companies rarely die from moving too fast, and they frequently die from moving too slowly.
>
> Reed Hastings, Netflix Founder and CEO (Hastings 2011)

Reed Hastings found himself in the middle of a customer and social media firestorm in the fall of 2011 when Netflix announced a plan to reinvent itself as a streaming media access provider—to the disadvantage of its DVD rental customers. Netflix intended to spin out its original DVDs-by-mail subscription business into a separate company called Qwikster while the Netflix brand would offer only streaming video options. The company's most popular $9.99 per month subscription plan for combined streaming plus DVDs-by-mail would be eliminated. Customers would have to choose between subscribing to DVD titles through Qwikster and accessing streaming media titles from Netflix—or be forced to pay monthly fees to each company to maintain both types of access. It seemed that Hastings had decided to test the boundaries of his theory that companies "rarely die from moving too fast" by pushing Netflix customers to immediately adopt his vision for the future of streaming media.

A significant portion of Netflix customers, many of whom had been loyal to the DVD rental service for over a decade, voiced their strong objections to the separation of Netflix into two stand-alone businesses. It was bad enough that they were they being asked to pay an extra monthly fee for maintaining a DVD subscription account along with access to Netflix's still-limited list of on demand streaming video titles. The positioning of the newly formed Qwikster as a backward-looking, DVD step child to Netflix's future growth strategy added insult to injury. It was not even clear how popular features such as customer ratings and the Netflix Cinematch recommendation system would be integrated between Quikster and Netflix after the split.

M.J. Cronin, *Top Down Innovation*, SpringerBriefs in Business, DOI 10.1007/978-3-319-03901-5_3, © The Author 2014

After a month of punishing customer attrition and plummeting market valuation, Hastings had to acknowledge that his bold vision had turned into an innovation nightmare. In October 2011 he issued a public apology to irate customers and announced that Netflix was abandoning the Qwikster spin-off and would keep DVD and video streaming as integrated Netflix services.

At a time of increased competition among providers of streaming video services and sustained pressure on DVD subscription pricing, Netflix had failed a self-imposed test of the strength of its customer loyalty and the barriers to exit in its customer ecosystem. Even though Hastings backpedaled on the Qwikster plan and apologized to his customers, the negative impact on Netflix lingered. Analysts downgraded Netflix stock, noting that the company's market lead in DVD inventory, content discovery and delivery services had taken years of capital intensive investment to build—and that much of the current Netflix infrastructure would not provide any competitive advantage in the streaming media sector. Netflix stock price declined from its high point of $300 per share in July 2011 to a low of $63 at the end of November of that year as investors lost faith in the company's ability to continue to grow its customer and revenue base. Almost a million customers, some of whom had maintained their DVD subscription and paid the monthly fee even though they were no longer frequent users, decided to cancel their subscriptions. In the last quarter of 2011, few observers were bullish on the future success of Netflix as the company scrambled to regain customer loyalty while recalibrating the timing of its transition away from the DVD format that had been the foundation of its success.

This case study will contrast the Netflix Nimble Pacesetter strategy that resulted in its market leadership in the DVD-by-mail category with the challenges it faces in achieving a sustainable competitive advantage through innovation in streaming media services.

Betting on DVD Technology, Process Innovation, and Data Mining

According to the often-repeated but apparently apocryphal Netflix founding story, in 1997 Netflix co-founder Reed Hastings was so aggravated at having to pay a $40 fine to Blockbuster for an overdue video that he vowed to start a business that would eliminate overdue video fines completely. As the story goes, Hastings experimented with mailing DVDs to himself and when they arrived in good shape he decided that it was feasible to rely on the US post office to lower the new company's fulfillment costs. His aggravation yielded a strategy for disrupting the prevailing storefront video rental model with its reliance on a small inventory of popular titles and stiff overdue charges. Over the next decade, Netflix created a thriving DVD-by-mail service, transforming the $11 billion dollar US video rental sector and driving arch-rival Blockbuster into bankruptcy.

According to Netflix co-founder Marc Randolph and other early Netflix staff members, Hastings actually crafted the story about his eureka moment years after the company founding and used it as a marketing pitch to summarize the core Netflix value proposition. The punch line of the founding story emphasized the value of convenient home delivery and return of DVD titles by mail, along with allowing customers to keep discs as long as they liked with no worries about late fines. In fact the original Netflix revenue model did not match up to Hastings' vision. The company started out charging a traditional time-limited rental fee ($4 for a 7 days rental) with a promised 2–3 days delivery time and a modest discount for renting four DVDs at a time (Thomas 2012).

Charging consumers a per-title rental fee and asking them to wait for several days before the disc arrived in the mail was not such a dramatic improvement over the Blockbuster model. And it came with a giant disadvantage—Netflix could not hope to cater to the impulse customers who waited until the last minute to decide that staying home and watching a video was their weekend activity of choice. Even though video rental growth had lost some momentum by the end of the 1990s, Blockbuster had established a network of thousands of storefront video outlets, making the brand a convenient drop-in location across the United States. By 1999 the company had captured 30 % of total US market share and had retooled its own business model to improve cash flow and title selection. Instead of paying film studios up to $80 per unit for each major new release, Blockbuster implemented a revenue share agreement with key studios in which it paid between $3 and $8 per unit plus a 40 % share of the rental revenues over time (Siklos 1999).

No wonder that Blockbuster, Hollywood Video and other storefront video rental companies did not regard the start up that specialized in renting DVDs by mail as much of a threat. Visitors who tried out the Netflix rental service in its early days had to work their way through a cluttered and non-intuitive website that offered little help in selecting a film that they might like to watch. Once they made their selection, customers then had to be patient while waiting for a disc to arrive. Since all DVDs were shipped from a central site in California, the further away from Netflix headquarters the customer lived, the longer it would take for delivery. With a postal mail delivery cycle that could take up to 5 days depending on the customers' location, and a price point that was similar to renting at a storefront a few blocks away, new Netflix users had few incentives to become regular customers or to drop their Blockbuster memberships.

Fortunately for the long-term survival of Netflix, the company did offer one distinctive value proposition—its focus on stocking DVD format films rather than video tapes. Netflix's online promise of providing "Virtually all DVD titles!" was not just a marketing ploy. Netflix bought up all available DVD film titles to become the clear leader in the nascent DVD entertainment sector. Even though DVDs were still an emerging movie format, Hastings bet that DVD technology would quickly overtake videotapes as the preferred film distribution and viewing mechanism. He proved to be prescient. Netflix opened for business at the moment when the market for home DVD players was taking off and winning adopters at a rapid pace. The installed base of DVD player hardware in the US surged from under 1 million players in 1998 to over 12 million players in 2000, fueling a demand for DVD film titles just at the time when Netflix was ramping up its inventory by investing in both mainstream titles and independent films that were not available from Blockbuster and other rental storefronts.

Nonetheless, Netflix in its early years was a company desperately in need of business, process and services innovation. It had to find a way to cut down on delivery time to its far-flung membership base and to generate more value from its growing title inventory. It needed a content discovery solution that could recommend the titles most likely to appeal to its customers—especially the older, foreign and less popular independent film titles that did not feature in mass media reviews and top ten lists. To improve profitability and cash flow, it needed to follow the example of Blockbuster and move from expensive per-title purchase agreements with film studios to revenue sharing models. Finally, Netflix had to make better use of all the data it was collecting about customer film ratings, content popularity and rental demand patterns to enable innovation across every aspect of its business.

It would take most of the next decade for the company to create the integrated solution of recommendation algorithms, data mining systems, logistics, distribution processes and partner–customer ecosystems that would transform Netflix from a startup in the new category of providing DVDs-by-mail to an innovative and disruptive Pacesetter.

Integrated Building Blocks

Netflix is the world's largest online DVD rental service, with more than 600,000 subscribers and a comprehensive library of over 11,500 titles. For $19.95 a month, Netflix subscribers can rent as many DVDs as they want, with up to three DVDs out at a time, and keep the movies for as long as they like. There are no due dates and no late fees. DVDs are delivered directly to the subscriber's address by first-class mail. Based in Los Gatos, California, the company also provides background information on DVD releases, including reviews, member reviews and ratings, and personalized movie recommendations. (Netflix Press Release 2002)

Among the first challenges that Netflix tackled was transforming its fulfillment processes to cut down on the time required for delivery of discs to its customers around the US and devise a way to put those discs back into circulation without delay. One basic innovation that helped to improve the speed of delivery was the custom-designed Netflix mailing and return envelope and sleeve. The inner sleeve that surrounded and protected the DVD from scratches was imprinted with a barcode that contained content and location information that could be scanned through a cut out window in the outer envelope. As a marketing bonus, the bright red envelope became an instantly recognizable icon of the Netflix brand.

The envelope was designed to comply with US postal requirements for first class mail delivery as well as to automate disc sorting using specialized Netflix equipment in the company's dedicated central DVD distribution center in San Jose, California. The central sorting facility could handle up to half a million DVD orders daily, but improving internal efficiencies at this location was not enough to solve the problem of speeding up delivery away from the west coast.

By 2002, Netflix customer base had expanded to every part of the US and the company decided to invest in a network of distribution centers. It announced the opening of its first 10 centers that year, launching a network that would expand to 58 highly automated distribution centers by 2011. Each distribution center was equipped with network-connected sorting equipment customized for the Netflix envelopes. Netflix staff worked around the clock, coordinating their processes closely with nearby US postal hubs to handle the delivery of hundreds of thousands of DVDs daily. Netflix even recruited the US Postal Service into its innovative processes with agreements to deliver returned discs directly to the nearest distribution center rather than handling them through the local post office. The goal of next day delivery and overnight return and processing of DVD's became a reality for most Netflix customers.

Custom envelopes, bar code scanning, decentralized distribution centers and automated sorting equipment might seem like obvious steps to take to streamline DVD delivery logistics—just as Amazon's one-click buying methods seemed obvious to other online retail stores after they saw the system in action. But these were innovations that enabled Netflix to speed up and simplify its logistics to offer disruptive home delivery speeds. Like Amazon, Netflix claimed patent protection for its innovations to prevent competitors from copying its ideas. Patents for the design of the delivery envelope and the overall process of distribution center management and integrated rental queue management were granted to Netflix in 2006, an opportune time for the company to assert its claim to these process innovations. As soon as the patents were granted, Netflix went to court to sue Blockbuster for infringement. In 2007 Blockbuster settled the infringement case, providing a clear signal that Netflix intended to protect its market lead by any means possible. The patent suit was one of the decisive blows against Blockbuster's belated entrance into the DVD delivery business.

In tandem with its logistics innovations, Netflix was working to expand its content inventory and to improve its recommendation system to match up customer preferences with the broad range of titles it had to offer—especially with more obscure independent films, older TV series and documentaries. Netflix addressed the issue of scaling up its inventory of popular titles by creating a new business partner ecosystem with the film studios, TV networks and independent content producers. In 2000, Netflix hired a director of content development who came from the film industry and led the effort to negotiate better terms for accessing content at scale. Over the next several years Netflix switched from purchasing the rights to each title separately to signing revenue-sharing agreements with the film studios and TV cable networks that set lower initial licensing fees for making multiple copies of the most popular titles available as needed in exchange for sharing revenue on the basis of the title's popularity with Netflix subscribers.

Progress on these fronts allowed Netflix to promote a more aggressive subscription model in which customers could borrow as many DVDs as they desired every month with the only limit being the number of discs that could be borrowed at any one time. By 2004 Netflix offered a deal that more closely matched Hastings' vision for industry transformation. Customers could sign up for a DVD subscription that cost under $20 per month and could have three titles on loan at any given time with no limits on the total number of discs per month as long as those borrowed previously were returned.

As Netflix fine-tuned its logistics to achieve a 1 day delivery and return timeline, the standard $19.95 monthly subscription fee would allow customers who returned their discs promptly to get new titles delivered almost every day. This was a breakthrough in service and content availability that trumped the convenience advantage of local video rental storefronts. In combination with the significantly larger title inventory and the Netflix patents, the evolving Netflix business model provided a considerable barrier to entry for would-be competitors. In addition, the change in customer expectations began to threaten the viability of the long-standing leaders in video rentals.

Consumer willingness to adopt a mail delivery service with no late fees forced the storefront video rental sector, including Blockbuster, to rethink its business model. Blockbuster had to abandon the hundreds of millions of dollars in profit that it had earned from all those annoying overdue fines. It began to close storefronts as customers migrated away to become Netflix subscribers. Blockbuster tried to fight back by imitating the Netflix subscription and delivery model. For a brief time Wal-Mart joined in the competition, announcing that it too was offering a subscription-based DVD rental service leveraging its storefronts for immediate distribution of the large inventory of DVD titles that it already offered for sale.

This competition was too little and too late. Netflix had already established itself as the leading DVD rental brand and it continued to scale up its subscribers, the depth of its title list, and its streamlined delivery service. Its core innovations in the distribution process realm were a deterrent to new entrants, who would have to create comparable DVD-tailored delivery capabilities to match the cost and volume processing efficiency of Netflix distribution centers.

Even as it bested the competition in process and delivery services, Netflix recognized that logistics and inventory management were not the only areas in which it had to innovate to maintain its pacesetter lead. One of the company's most important assets was the data it was amassing about customer behavior and viewing preferences. To get more value from this data, it was essential for Netflix to create a recommendation engine that could accurately suggest film titles that customers might enjoy based on their past viewing decisions and film ratings.

Netflix developed a proprietary collaborative filtering recommendation system called Cinematch to accomplish this goal. The algorithm that powered Cinematch used data from the five-star rating system that each subscriber was asked to complete for every DVD borrowed, plus additional subscriber reviews and recommendations to friends. This algorithm became a very effective means for Netflix to manage its inventory as well as to increase customer satisfaction and loyalty. The Cinematch recommendation service directly improved the company's profitability by encouraging customers to borrow lesser known titles that were not covered by revenue sharing agreements with the film studios. The more it could recommend such films while still providing its customers with a viewing experience they enjoyed, the more value and profitability Netflix could extract from its inventory and the easier it was to manage availability of popular titles that were in high demand immediately after their release or after winning an Oscar.

In 2007 Netflix marked the milestone of having delivered one billion videos to its customers. An IPO in the same year validated its market leadership. During the same period Netflix continued working to improve its Cinematch system, having seen the accuracy of its recommendations (as measured by the response of customers in ordering recommended titles) plateau at the 75 % level. Hastings suggested that Netflix sponsor a contest to get outside input on how to improve the algorithm results. To motivate participation in the contest, Netflix offered a $1 million cash prize to the individual or team that could improve the Cinematch algorithm recommendation system by at least 10 % based on Netflix data and criteria.

It was not the first time that a company had offered a prize for innovation or outsourced its data mining and analytics requirements but nonetheless the very public nature of the Netflix challenge captured the attention of the media as well as of developers. The Netflix Prize competition attracted over 30,000 teams from 170 countries and garnered considerable free publicity for Netflix. It turned out to be a very hard technical challenge to improve on the internal Netflix recommendation results. Over the next 3 years Netflix reviewed the contest submissions and regularly announced that no entrants had met the improvement standard set for the award. Finally in 2010 the Netflix Prize was awarded to a coalition of four teams from four countries—prior rivals who pooled their ideas to come up with a winning entry. In the meantime, according to the terms of the contest, Netflix was entitled to utilize the ideas from all entries to continue its internal work in improving Cinematch effectiveness.

Despite the success of the Netflix DVD-by-mail model and the competitive advantages that the company had created in its first decade, by 2010 Hastings was determined to make the transition away from DVDs to focus on streaming media. At this point, the installed base of DVD players in the US was close to 300 million units and more than 10 billion DVDs had been shipped to US consumers since 1997. The growth curve of DVD adoption had flattened out, but the millions of DVD owners were still committed to the DVD format for home entertainment. What Hastings overlooked in his eagerness to cut Netflix ties to a format that would eventually be replaced by online and mobile streaming was that regular DVD users would not change their habits overnight. Many customers wanted the option to keep their Netflix subscriptions in both DVD and streaming formats during a multiyear period of transition.

Rather than risking the downward spiral of the Innovators Dilemma, Hastings announced his plan to Netflix customers in fall 2011. Although his timing was premature, his vision for the future of the industry was well grounded. By 2013 Netflix had recovered from the 2011 debacle and resumed its rapid growth, adding subscribers in 40 countries for a total of almost 38 million members. Its stock price as of September 6, 2013 had recovered to $297.50, close to its July 2011 high point. As Hastings had anticipated, streaming has become the preferred means of accessing entertainment across a proliferation of mobile-internet connected devices as well as on home TV sets. DVD-by-mail subscriptions were declining rapidly (partly because Netflix no longer promoted them as a preferred membership option). Now Hastings faced a new challenge—could Netflix make a nimble transition to become a pacesetter in the highly competitive landscape of Internet TV and mobile streaming media?

Netflix Future

> Netflix is the world's leading Internet television network with nearly 38 million members in 40 countries enjoying more than one billion hours of TV shows and movies per month, including original series. For one low monthly price, Netflix members can watch as much as they want, anytime, anywhere, on nearly any Internet-connected screen. Members can play, pause and resume watching, all without commercials or commitments. (Netflix 2013)

Netflix has stopped describing itself as "the world's largest online DVD rental service," in favor of a more ambitious characterization as "the world's leading Internet television network." Hastings is once again attempting to disrupt industry leaders with Netflix business and technology innovations. This time around, however, he faces a much more complicated competitive landscape. The traditional industry players in film and TV production have already been joined by cable companies, wireless carriers and Internet service providers as well as a growing list of ecommerce and IT giants including Amazon, Apple, Google and Microsoft who are pursuing their own Internet TV and streaming media agendas. Instead of being a first mover and category creator as it was in DVD-by-mail service, Netflix is contending with a pack of deep pocketed rivals in a fight for digital home entertainment dominance that has been underway for more than a decade. Can Netflix apply the innovations that served it so well in the past, or does it need a different strategy to take on the streaming media marketplace?

To answer that question, it helps to consider some of the key challenges to becoming a world-class Internet TV network—including content acquisition and selection, proprietary vs. open hardware and software access platforms, quality and reliability of streaming media delivery to the home TV and to a wide variety of mobile devices.

Streaming Content Challenges and Innovations

While the number of titles available for streaming by Netflix has steadily increased, acquiring content has been a difficult and expensive challenge. Under US law, physical media formats such as books, music CDs and video DVDs, can be loaned out by the original owner. This "First Sale Doctrine" is what protects the rights of libraries to loan books and media to their users. As long as Netflix purchased the physical DVD (or licensed the right to make multiple copies of a DVD title) it could expand its inventory to include all the DVDs that are on the market. Over the years, Netflix has done exactly that, increasing the size of its inventory to cover most new media releases and many reissued older films and TV shows (Gallaugher 2013).

Digital distribution rights are not covered by the First Sale Doctrine and content owners are far more reluctant to license their content for unlimited online streaming. Film studios and TV producers are determined to make as much revenue as possible on every licensing deal. Streaming rights are usually licensed for a limited time period and costs often rise when licenses are renewed. Netflix financials indicate this trend, with the expense of acquiring streaming content rising from $48 million in 2008 to over a billion in new licensing deals in 2010, including a $900 million agreement for online rights for 3,000 films from Viacom, Lions Gate, Paramount and MGM. But even that high priced deal in 2010 only raised Netflix's total streaming content inventory to just over 20,000 films and TV shows—in stark contrast to the deep inventory of over 100,000 DVD titles that the company had built up during its first decade (Smith 2011).

Complicating content licensing even further, now that Netflix is defining itself as a direct competitor to the media producers and content owners, there is an increasing risk that some content owners will refuse to do business with Netflix at any price. As license agreements expire, and content owners make new exclusive deals with Netflix competitors such as Amazon, subscribers are complaining that some films and TV shows they want to watch are no longer available through Netflix.

Netflix can counter the risk of being shut out of content licensing deals by developing and producing its own exclusive content. Even though network quality content production is itself a risky and expensive strategy, Hastings has decided to pursue that option. Netflix paid $100 million for the initial 26-episode exclusive for the series *House of Cards*, beating out HBO and AMC in a bidding war for a brand-new series featuring Oscar-winner Kevin Spacey. The series was well received by consumers and critics, garnering an Emmy award. Netflix has increased its commitment original content creation, airing more new series in 2013 and signing a deal with Sony for 2014 content production.

Despite these investments in content creation and licensing, it seems unlikely that Netflix will be able to match the content inventory advantage that it enjoyed in its DVD business. This has repercussions for the value of Netflix innovations in collaborative filtering and the Cinematch recommendation system. The lack of a "long tail" of titles to match up to customer preferences reduces the value of the recommendation system since there are fewer undiscovered titles available to recommend. On the plus side, streaming video services provide Netflix with another source of data about customer preferences and viewing habits. Since subscribers must be Internet connected to watch content, Netflix now knows exactly when, how long and how often its customers are interacting with its content. It has developed new algorithms to balance the server and streaming loads by matching viewers streaming requests to the cloud server location able to deliver the bits most efficiently, one of the innovations that Netflix is counting on to address its intensive use of bandwidth during prime time viewing hours.

Streaming media providers need a platform to link their content to home TV and Internet connected devices such as smartphones and tablets. Traditionally, the preferred platform was a branded hardware device (such as the cable set top box, Microsoft's Xbox, Apple TV and most recently Google's Chromecast device) that connected directly to the TV set. While developing such a device is not technically difficult, that is only step one. Convincing consumers to add a new gadget to the already complicated home entertainment set up has been a high barrier to entry.

Despite the stiff competition, Hastings was attracted to the advantages of introducing a Netflix branded streaming device and an internal team of Netflix engineers developed a hardware prototype. As the number of competing devices multiplied, however, Hastings was persuaded that a branded device would be seen as competitive, reducing the chances for Netflix to partner with other device makers. The company decided to drop the branded box in favor of developing a Netflix software interface that would be compatible with as many hardware devices as possible. The internal device prototyping team spun out of Netflix to become Roku, offering a standalone streaming device for $99. It was just the start of a large partner ecosystem that has expanded to include hundreds of devices from leading consumer electronics vendors such as Sony, Panasonic, LG, Samsung, Toshiba, Microsoft and Apple and well as start up device makers. Netflix apps currently stream content to smartphones, tablets and game consoles as well as to TV sets giving it the broadest possible market for its streaming services. This success is not unique to Netflix, however. Its competitors (even those like Amazon that also have branded devices for media consumption) have adopted a similar platform and mobile app strategy and created their own extensive partner and distribution ecosystems.

The Netflix DVD delivery service leveraged the US post office home delivery service, a cost-saving decision that allowed the company to invest in logistics innovations that linked local Netflix distribution centers to postal processing hubs to improve delivery and turnaround times. The Post Office was a willing partner because Netflix shared the labor of sorting and processing while paying the cost of first class postage.

Netflix relationships with the companies that provide consumer Internet service and broadband connectivity have not been so harmonious. Netflix video streaming is widely seen as using a disproportionate share of bandwidth without paying its share of the cost of home delivery. There is no disputing that Netflix customers use an enormous amount of bandwidth watching their favorite media content. Netflix streaming video services currently account for about 30 % for total North American Internet traffic on a daily basis. Cable companies, wireless carriers and Internet Service Providers (ISPs) are actively seeking ways to offload some of the costs of providing the bandwidth that supports all that streaming entertainment back to Netflix and its subscribers. So far, the Federal Communications Commission has upheld a doctrine called Net Neutrality which prohibits regulated connectivity providers from taking measures to throttle back the bandwidth available to streaming video thus protecting high speed Netflix access to its customers' homes and mobile devices. However, this battle is far from over as Internet providers regularly challenge Net Neutrality in court. If the Internet providers prevail, Netflix may face degradation in the quality of its streaming content to reduce the burden on bandwidth, or a data surcharge, or both (Knutson and Ramachandran 2013).

A related challenge is the reliability and control of streaming services. Netflix contracts with Amazon's Web Services and Amazon public cloud infrastructure to host its media content library. Although the Amazon cloud is managed separately from Amazon's competitive media streaming business, the need for Netflix to rely on a direct competitor for an essential service highlights the complexity of the competitive landscape. In the world of streaming media infrastructure and delivery, there is no Netflix-controlled analog to the custom-designed and patented Netflix DVD delivery envelope.

Conclusion

The combination of steep streaming media licensing fees, high costs for exclusive original content, and subscription price reductions to compete with competitive options such as the "free" Amazon Prime streaming access plan mean that Netflix has transitioned from a cash cow, high-margin business in DVDs to a high-risk venture in streaming media with, at best, lean and less predictable profit margins for the next several years. As of October 2013, Netflix is back in favor with consumers and financial analysts. Its US paid subscriber base has topped 30 million, putting it ahead of HBO—the competition for its new goal to become the world's leading television network. Reed Hastings decision to move fast enough to avoid being caught in the Innovators Dilemma has paid off, at least in the short term. As the chief visionary for Netflix, Hastings is busy connecting the building blocks and data insights from streaming media to replicate its Pacesetter position in the larger and more competitive Internet TV sector.

References

Gallaugher, J.: Information Systems: A Manager's Guide to Harnessing Technology. Flat World Knowledge, Boston (2013)

Hastings, R.: An explanation and some reflections. Netflix Blog Posting (September 18, 2011). URL: blog.netflix.com/2011/09/explanation-and-some-reflections.html

Knutson, R., Ramachandran, S.: Verizon fights FCC on web rule Wall Street Journal, Business (September 8, 2013). URL: online.wsj.com/article/SB10001424127887323864604579063203354574082.html?mod=WSJhpssectionsbusiness

Netflix: Press release. Netflix (June, 2002). URL: www.netflix.com

Netflix: Netflix investor relations overview. Netflix (September 1, 2013). URL: www.netflix.com

Siklos, R.: Blockbuster finally gets it right. Business Week (March 7, 1999). URL: www.businessweek.com/stories/1999-03-07/blockbuster-finally-gets-it-right

Smith, E.: Netflix CEO unbowed. Wall Street Journal (September 20, 2011). URL: online.wsj.com/article/SB10001424053111903374004576581000189433470.html

Thomas, O.: Netflix's forgotten cofounder says Reed Hastings is lying about how he came up with the idea for the company. Business Insider (October 4, 2012). URL: www.businessinsider.com/netflix-forgotten-cofounder-marc-randolph-2012-10

Chapter 4
Nokia Drops the Torch

How did we get to this point? Why did we fall behind when the world around us evolved? This is what I have been trying to understand. I believe at least some of it has been due to our attitude inside Nokia. We poured gasoline on our own burning platform. I believe we have lacked accountability and leadership to align and direct the company through these disruptive times. We had a series of misses. We haven't been delivering innovation fast enough. We're not collaborating internally. Nokia, our platform is burning.
  Stephen Elop, Nokia President and CEO (Elop 2011)

When Stephen Elop joined Nokia as President and CEO in September 2010, he faced a daunting turnaround challenge. Nokia's mobile device market share, which had peaked in 2007 at around 40 % worldwide, had fallen to under 30 %. The decline coincided with the launch of Apple's wildly popular iPhone line and Google's release of its open and free Android mobile operating system. This new competition triggered a tectonic shift in consumer demand, wireless carrier and app developer interest away from Nokia's classic hardware-centric phone designs to Internet-connected smartphones with an abundance of applications. But that long-term shift and Nokia's inability to keep pace are only obvious in retrospect. Elop joined Nokia amid expressions of optimism from the company's directors that "The core strategy is solid and Nokia will continue to power through what is a substantial transformation (from a hardware company to a software company). Elop will help to accelerate that" (Nokia 2010).

With hindsight, we know that 2010 marked the midpoint of Nokia's 6 year slide from a mobile market pacesetter to a 2013 also-ran. Less than a year after taking the helm to spearhead the expected turnaround, Elop was plaintively asking for explanations of Nokia's accelerating decline. How had a challenging but salvageable situation turned into an outright market debacle in such a short time? Stephen Elop's frequently quoted "Burning Platform" memo to Nokia staff didn't do much to illuminate the deeper problems his company was facing, any more than it served to rally his employees or convince them to support his vision for Nokia's future.

Elop's 2011 decision to abandon Nokia's operating system and its patent-generating in-house research and development efforts in favor of a do or die partner-ship deal with Microsoft Windows Mobile won even less favor in the marketplace.

M.J. Cronin, *Top Down Innovation*, SpringerBriefs in Business,
DOI 10.1007/978-3-319-03901-5_4, © The Author 2014

The negative trends that Nokia's CEO described so graphically in his 2011 speech worsened in the 2 years that followed. Nokia cut back on staff, closed manufacturing plants and slashed operating expenses but did not manage to reverse the decline of its market share. The company that had dominated the mobile device landscape for over a decade lost its global lead to Apple's iPhone and the manufacturers that adopted Google's Android mobile operating system. In summer 2013, Nokia suffered the humiliation of falling behind Samsung's share of the mobile phone market even in its native Finland. In September 2013, Microsoft announced that it would acquire Nokia's mobile device business for $7 billion, putting an end to Nokia's long run as a mobile device Pacesetter.

As Stephen Elop asked, but failed to answer adequately in 2011, why did Nokia fall behind its competitors after building such an impressive track record as a market leader in the previous decade?

This case study analyzes how Nokia rose to become an Innovative Pacesetter for mobile devices between 2000 and 2007 and why it stumbled so quickly when the mobile industry's competitive landscape and innovation challenges shifted. It discusses the company's rise and fall in the context of critical flaws in Nokia's innovation strategy during the past decade.

2010 Pivot Point: Enter a New Race or Stay the Course?

The analysis starts with a brief look at Nokia's top level management understanding of the mobile market challenges and options facing the company at a critical juncture in 2010.

A few months prior to Elop's hire, Anssi Vanjoki, one of Nokia's top managers, acknowledged the company's market share decline in the same breath as declaring that Nokia was well positioned to regain global leadership through internal improvements in quality, processes and product development. Vanjoki's July 2010 posting to Nokia's corporate blog emphasized advancing current projects rather than rethinking overall strategy, indicating that Nokia executives were unwilling to confront the widening gap between Nokia products and consumer preferences—at least in public.

> As head of Mobile Solutions, it's my aim to ensure Nokia stays as the market and intellectual leader in creating the digital world. I'm under no illusions; it's no small task. Over the coming months, we'll be advancing current projects and working to simplify the way we work in order to deliver products and services faster, and with a laser focus on quality…
> I am committed, perhaps even obsessed, with getting Nokia back to being number one in high-end devices. … We have all the assets — including R&D and product development – at our disposal under one roof – to produce killer smartphones and market-changing mobile computers. (Vanjoki 2010)

In the same 2010 blog post, Vanjoki asserted that Nokia was confident that it could recapture market share while sticking with internal mobile device designs and software platforms, specifically the aging Symbian mobile operating system.

This was a significant decision at a time when Google's Android operating system was becoming the platform of choice for mobile phone makers around the world. A move from the widely criticized Symbian legacy system to the up and coming Android open platform would have signaled that Nokia understood the urgent need to reinvent its developer ecosystem and partnering strategy. Vanjoki and other top Nokia executives went out of their way to assert the opposite. "We have no plans to use any other software. Despite rumors to the contrary, there are no plans to introduce an Android device from Nokia. Symbian is our platform of choice for Nokia smartphones," Vanjoki concluded in his "fight back" posting.

A year earlier, Vanjoki had discounted the competitive threat from Apple's iPhone, predicting in an interview that after the initial market buzz the iPhone would settle into a small niche in the same way that Apple's Mac line of PCs appealed to a fraction of the global market compared with the dominant Microsoft Windows market share. Nokia's underestimation of the iPhone's market appeal was echoed by the company's engineers who conducted a series of tests on Apple's new product and reported that it failed many of Nokia's standard performance tests including surviving a drop onto a concrete floor (Mac Daily News 2009).

Such comments reflected a strongly held belief among managers and technical staff that Nokia's market downturn was a temporary setback, a blip in the company's decade-long track record as a mobile innovator. To supplement its internal R&D, Nokia had acquired a number of innovative companies specializing in high growth areas such as social media, photo sharing (Twango), mobile advertising (Enpocket), mobile search (Metacarta) and web browsing (Novarra). It spent over $8 billion in 2007 to buy Navteq, a digital mapping specialist to shore up its software and services portfolio. Such acquisitions helped to justify top management's public optimistic outlook. Expectations of a turnaround were further bolstered by improvement in Nokia's financials for the first half of 2010, which reflected increased revenue and respectable profit. Even more encouraging, Nokia's latest smartphone devices were selling reasonably well.

Despite its attempts at reinvention through acquisition, Nokia was suffering from what Clayton Christensen has dubbed the Innovator's Dilemma; its success in the first mobile market boom had blinded Nokia's leaders to the severity of threats from new entrants with very different technologies and business strategies. As Stephen Elop was soon to discover, there was little internal appetite for a radical change of course at Nokia.

Setting the Pace for a Generation of Mobile Phones

Founded in 1865 in Finland as a paper mill operator, Nokia demonstrated an impressive ability to shift course and adapt to new market opportunities during its first 100 years of operation. The company pioneered the application of emerging technologies in manufacturing and moved into industrial rubber goods in 1898 when that sector began to expand dramatically with new extrusion technologies and global demand.

It started manufacturing communication cables in 1912 and enjoyed decades of steady growth in as a diversified industrial manufacturer.

In 1963 Nokia moved into radio telecommunications, television and electronics manufacturing and these sectors became its primary growth drivers for the next several decades. The company marked some notable innovation milestones in the early days of mobile telephony, rolling out one of the first car phones in 1984 and a portable mobile handset in 1987. These early devices were too cumbersome and expensive for the average consumer and the limitations of analog network connections subjected the early phones to interference and dropped calls. Another barrier to adoption was the lack of any unified communications standards supported by a global market of wireless carriers. Nonetheless, its focus on mobile devices enabled Nokia to master phone manufacturing processes and to assess the growth potential for mobile.

GSM (global system for mobile communications), the European response to these barriers to growth, had been geminating for most of the 1980s starting with a working group to develop a European standard for digital cellular telephony. Nokia played a major role in developing the new standards throughout the decade. When the EU agreed in 1987 to make GSM a mandatory standard for all European cellular networks, Nokia had a competitive head start in the race to design and develop new GSM devices and equipment. As the company's web site describes this turning point, "In 1987, GSM (global system for mobile communications) is adopted as the European standard for digital mobile technology. With its high-quality voice calls, international roaming and support for text messages, GSM ignites a global mobile revolution. As a key player in this new technology, Nokia is able to take full advantage" (Nokia 2013).

Nokia made a strategic business decision to bet its future on the growth of GSM technology. By 1991, the company had divested its other manufacturing divisions to focus exclusively on manufacturing mobile phones and GSM telephony network equipment. The company's engineering and technical staff assumed leadership roles in key standard setting committees within the European Telecommunications Standards Institute (ETSI), the group responsible for further development of the GSM cellular standards and device specifications that became the basis of Europe's rapid adoption of mobile phones in the next decade.

Nokia's decision to develop GSM devices and its active participation in setting the network and mobile device standards ensured that Nokia equipment found a place in many of the cellular telephony milestones of the 1990s. When the first GSM call was made in Finland in 1991, it was transmitted using Nokia networking equipment. By 1992, Nokia had launched its first GSM mobile phone, the Nokia 1011. It was the first of a series of models based on the Nokia manufacturing template that came to be known as the "candy bar" design featuring a prominent dial pad and a small display screen.

Twenty years after their market launch, the early Nokia phones seem closer to the traditional landline handsets that predated them than to the full color touch screens of today's smartphones. The most innovative feature of the Nokia 2100 was its musical ring tone. The 2100 succeeded because mobile phones had finally become affordable and, thanks to the GSM standards, reliable. Consumer demand soared in Europe and everywhere around the world. The Nokia 2100 series phone eventually sold 20 million units globally making it the most popular mobile device of its era.

Just as with Netflix and mass market adoption of DVD players and discs, Nokia growth was fueled by the global rise in GSM networks and mobile subscribers. In 1994 there were 100 operator members of the GSM Association, expanding to over 200 operators in 100 countries by 1997. The user base for GSM phones sky-rocketed from 1 million GSM subscribers in 1994 to over 100 million subscribers by 1998—a 100-fold increase in the available consumer market opportunity for Nokia phones. By 2001 the number of GSM subscribers had reached 500 million, with no slowdown in sight. In 2005, Nokia celebrated the sale of its billionth mobile phone. That same year, the number of global mobile subscribers topped two billion. In the next few years, Nokia would achieve its pinnacle of market share and corporate value. It had succeeded in staying on top of the GSM rocket ship for over a decade.

Nokia, flush with year after year of rapidly rising revenues and market share, spent lavishly on R&D to develop advanced wireless capabilities and prototype devices that showcased its technology. In 1996, the company launched the Nokia 9000 Communicator, a cellular rival to the then-popular Personal Digital Assistants (PDA) devices like the Palm Pilot and announced breakthroughs in multimedia mobile devices. In subsequent years, its engineers developed mobile gaming applications, e-health services, voice recognition, and mobile web browsing, mobile chat and integrated cameras. In 2002, Nokia R&D was working on touch sensitive screen technologies, and an app that displayed the names of songs and artists on screen while they were performing.

As documented by Nokia's history of R&D, Nokia labs produced a steady stream of technical breakthroughs, generating an impressive portfolio of 40,000 issued patents. The company's inventions and advances in mobile technology augured well for Nokia' long term outlook as a mobile pacesetter. Unfortunately, these innovations did not reflect the priorities of Nokia's primary customers, the wireless carriers. GSM carriers wanted incremental improvements to Nokia's previous phones—and they wanted to keep prices low to attract new wireless subscribers. Nokia's investors applauded announcements of technology breakthroughs but they were focused on the company's quarterly revenue growth and profitability rather than on unproven new product concepts.

Nokia's executives pleased their primary customers and responded to their investor demands by perfecting manufacturing processes to turn out reliable, contemporary looking handsets with high profit margins. Nokia executives could point with pride to advanced mobile technology that had been invented in the company's research labs, but most of these innovations didn't make it into commercial production. Nokia's annual growth was dependent on the rapid rise in market demand for GSM mobile phones and the expansion of GSM networks worldwide. Ironically, Nokia's success in pleasing its GSM carrier customers pushed it further down the path of the Innovators Dilemma.

Manufacturing Innovations: Efficiency Optimization

Nokia became a master of mobile manufacturing efficiency to meet the escalating global demand for mobile devices. By 2006 it was operating nine manufacturing plants around the world and producing over 325 million handsets every year. An article in *BusinessWeek* in 2006 noted that Nokia factories were manufacturing "10 mobile phones per second every hour of every day, all year long" (Reinhardt 2006).

Mobile phones are complex devices requiring multiple radios, chips and other electronic components. The assembly of an increasing number of different phone models required Nokia had to handle more than 100 billion parts in its factories. The company's long history of electronics manufacturing and its status as one of Finland's largest enterprises allowed it to attract well trained engineering and assembly staff. It worked with vendor partners and developed internal systems to coordinate the delivery of complex phone components with the requirements of its just-in-time assembly schedules. It relied on automation and internal process innovation to keep costs low and quality high, keeping control of all the manufacturing and assembly work in its own factories.

The company's manufacturing innovations, combined with an intense focus on efficiency, succeeded in driving down the cost of the assembly process and steadily increasing profit margins as Nokia's production soared. In 2006 it cost Nokia about 69 Euros in parts and labor to make its most popular model phones, devices that on average would sell for 102 Euros. Not only had Nokia become the world's largest phone manufacturer, it had achieved the highest gross margin in the industry.

Advanced manufacturing capabilities also enabled Nokia to meet the demands of the hundreds of wireless GSM carriers for constant phone upgrades and customization. Each wireless carrier had specific priorities for the mobile phone features that it wanted to make available to its subscriber base. Many carriers did not want to enable their subscribers to use advanced network features and alternatives to cellular connections such as Bluetooth, Wi-Fi and GPS. To the extent that they did enable these innovations on their network they insisted on controlling the ways in which consumers were able to access them by demanding that manufacturers turn specific options off at the factory. This resulted in a fragmentation in the features of phones offered by different carriers, even if the manufactured device was designed to meet all network standards. As the gateway to the market, the wireless carrier had the final word on phone features. A carrier could simply refuse to sell and to activate a particular phone model if that phone did not meet its unique specifications.

To keep gaining market share, Nokia was highly motivated to meet the requirements of its carrier customers whether these involved tweaking features of a particular phone model or rolling out an entirely new line of phones. The more phone models that Nokia was able to manufacture in a cost-effective way the more it could fulfill carrier needs worldwide and cement its lead over rival phone makers. Smaller manufacturers were hard-pressed to match the economies of scale and centralization or the process efficiencies that Nokia could command as market leader.

Even during this peak of company success and profitability, there were indications that Nokia's focus on manufacturing process efficiency was stifling its flexibility in device design as well as its longer term innovation capabilities. For example, the company was slow to adopt the very popular folding clamshell designs that Motorola and other manufacturers turned into consumer favorites. The familiar Nokia candy-bar form factor had started to look old-fashioned next to flip phones with larger internal display screens. But after optimizing its assembly processes and component calculations for the candy bar devices, any major new phone design would mean temporary hits to manufacturing efficiency.

Eventually as clamshell style phones captured a larger percentage of the market and wireless carriers began to request that Nokia offer this design, the company did bring them into production. But they had waited too long. By the time Nokia's clamshell models hit the market, the clamshell design was about to be vanquished by Apple's iPhone touch screen and sleeker, more forward looking form factors.

Challenges of Serving Wireless Carrier Customers

Nokia's market position in 2006 and 2007 reflected many elements of the classic Innovators Dilemma as described by Clayton Christensen. The company had achieved market leadership in mobile phones, in particular GSM network compatible mobile phones. The rapid growth of GSM wireless carriers and mobile subscribers worldwide had created a seemingly insatiable demand. Too keep up with this demand, Nokia focused on designing its mobile devices in ways that would optimize manufacturing efficiencies even when fulfilling numerous customization requests from its wireless carrier customers. The company became more and more efficient at manufacturing, reducing costs to achieve the highest gross margin in the industry.

Nokia also maintained a very high profile in the standards-setting groups such as ETSI, ensuring that its engineers and managers played an active role in establishing new wireless and device specifications. Nokia participated in the years-long discussions leading up to each advance in the design of core wireless networks and telephony features as well as every new feature that was approved for mobile devices. This intense activity in standards-setting groups gave Nokia a chance to influence the technical capabilities of each new generation of wireless communication. It also aligned Nokia even more closely with its GSM carrier customers.

Being on the inside track and partnering with the carriers to craft global standards for wireless had significant advantages for Nokia. For example, the company got a head start in developing the third generation (3G) phones designed to work on the higher speed 3G networks. Nokia's first 3G phone launched in 2002, giving the company an early start in provisioning higher speed mobile phones that could support more data and applications.

On the other hand, its commitment to supporting wireless carriers also led Nokia to abandon innovative breakthroughs that were not appealing to the carrier mindset. Carriers, for example, did not favor innovations that allowed subscribers to bypass paid cellular networks to communicate or download data over Wi-Fi and NFC (Near Field Communication). In the mid 2000s, carriers feared that widespread consumer use of these unregulated networks would reduce their average revenue per user, making it hard for the carriers to recoup the enormous capital investments that had been required to upgrade to 3G networks. Even though Nokia engineers developed NFC and Wi-Fi enabled prototypes, the company pulled back from their early launches of such devices in deference to carrier preferences.

The same preference for controlling their subscribers led to many wireless carriers to assume a gatekeeper role in allowing third party applications on their phones. Once Nokia and other manufacturers developed Java-enabled phones it was expected that consumer demand for apps would soar, creating a thriving app economy with developer ecosystems and app stores. Nokia started its own app store, but it did not push the envelope on app ecosystems and independent app store innovations in part because these were seen as competitive with carrier-controlled mobile application stores.

An unfortunate side effect of Nokia's reliance on wireless carrier channels was that it cut the company off from important data about how their phones were actually being used after purchase. In contrast to Amazon's constant analysis of how shoppers used its web site, and Apple's access to the data from its AppStore, Nokia had few direct connections with mass market mobile subscribers and no opportunity to build a sophisticated program of data collection and analytics based on subscriber behavior over time. Without such user data to supplement its internal forecasts of trends in mobile data adoption, Nokia misread the impact of third party mobile applications in shaping consumer preferences.

Nokia's decisions made good sense in the short term context of serving the interests of its most important and demanding customers—the wireless carriers. But as the Innovators Dilemma makes clear, long-term customers tend to dislike disruptive new technology innovations and may fear innovations that could transform their current market. As discussed earlier, disruptive technologies often do not perform as well as market leading products when first introduced. This performance gap makes it easy for established market leaders to underestimate the future impact of disruptive technology and extraordinarily difficult for them to pivot toward the disruptive products and services that will alienate their most profitable customers.

Such was the situation Nokia faced in 2007 when its managers and engineers discounted the importance of the iPhone launch and the Android operating system for the mobile competitive landscape. Even after its market share had taken a hit, Nokia in 2010 decided to stay on course and redouble its efforts to win back wireless carriers rather than launching forward-looking disruptive innovations.

Ecosystem Partner Relationships

Thanks to its involvement in 3G standards groups, Nokia was well aware that new handset technologies, including new operating systems, would be needed to leverage the emerging higher-speed, data capable networks. Popular PDAs (Personal Digital Assistants) such as the Palm Pilot provided one model for the design of advanced phone operating systems (OS). Nokia evaluated existing PDA systems and identified Psion Computers, a PDA hardware and software company, as a likely platform its next phone OS. Rather than buying the company outright, Nokia wanted to promote broader industry adoption of the platform. In 1998 it spearheaded the creation of a Joint Venture to further develop and license the Psion operating platform. In addition to Nokia, phone makers Ericsson, Motorola and Matsushita (Panasonic) and Psion Computers joined the consortium. Psion brought EPOC, its latest PDA operating system, into the group and the new OS was renamed Symbian.

Nokia's role in the Symbian Consortium illustrates how Nokia's dominant market position caused long term problems for its ecosystem and platform strategies. Since Nokia was by far the largest licenser (four to one over the other Symbian phone manufacturers) it leaned on the Symbian group to prioritize features and development that supported Nokia's device road map. In addition to participating in Symbian, Nokia continued to invest heavily in its own R&D and patenting programs. Rather than sharing its technical innovations with Symbian so that all consortium members would bring advanced devices to market, Nokia started undercutting its JV partners. A Symbian engineer recalls that in 2001 and early 2003 Nokia argued inside the consortium that it was premature to invest any resources on integrating camera phone capabilities into the Symbian platform. In 2002, Nokia announced the launch of its proprietary built-in camera phone model, jumping ahead of its partners. Similarly, Nokia insisted on using its own multimedia messaging specifications for picture messaging rather than supporting a standardized Symbian solution (Ocock 2010).

By 2007 application developers trying to tailor their apps to work on as many mobile phones as possible were frustrated with the limits of the Symbian OS. With its roots in the stand-alone PDA world, Symbian was not designed with Internet access and browser-based use of data in mind. It would seem that the very large base of Nokia phones worldwide would be an attractive market for app developers and ecosystem partners. However, the pattern of carrier customization meant that there was not typically a critical mass of one phone model that behaved the same way around the world.

From a developer standpoint, the fragmentation of the Symbian platform across hundreds of different Nokia phone models (thanks to that carrier insistence on customization) had become a nightmare. Even when developers were porting their apps to a single Nokia phone model, it was not likely that the apps would work identically on different carrier networks. Each app had to be tested separately for each network where the developer hoped to sell it. The costly overhead of testing and tweaking their apps for hundreds of variations of Symbian phones just wasn't worth it for most developers, especially after the iPhone AppStore and Android Marketplace entered the market.

More than one Symbian ecosystem partner blamed the failure of the OS on the carrier requirements for device customization that placed an impossible burden on app developers and phone manufacturers who had not mastered their assembly processes at the same level as Nokia,

> I remember in one case there were 10,000 requirements to get Symbian products onto that one carrier's network. A typical carrier requirement would be anything from do or don't include wi-fi support to where things showed up on a menu. (Best 2013)

In a partial response to Google's open source Android OS and to end years of dissention among the founding partners, Nokia bought out the remaining the Symbian JV owners in 2008 and announced the availability of an open source Symbian platform that would be managed by a non-profit Symbian Foundation. By that point, however, Symbian's reputation as a complex and outdated development environment had eclipsed the appeal of open source availability—especially in the context of growing manufacturer interest in Android. The Foundation failed to attract enough new participants and in 2010 Nokia announced that it was absorbing Symbian back into its organization, to be followed by the 2011 announcement that Nokia was dropping Symbian altogether in favor of the Microsoft Windows operating system.

After Nokia had abandoned or undercut the ecosystem strategies of the previous decade, Elop expressed a belated realization that thriving mobile software and application ecosystems were an essential source of innovation and competitive advantage.

> The battle of devices has now become a war of ecosystems, where ecosystems include not only the hardware and software of the device, but developers, applications, ecommerce, advertising, search, social applications, location-based services, unified communications and many other things. Our competitors aren't taking our market share with devices; they are taking our market share with an entire ecosystem. This means we're going to have to decide how we either build, catalyse or join an ecosystem. (Elop 2011)

Nokia's Pacesetter Dilemma

A company with Nokia's track record of invention can hardly be accused of having failed to innovate. In the last decade, Nokia has spent over $40 billion dollars in mobile R&D, more mobile research funding than Apple and Google combined. It has received patents in over 10,000 fundamental areas of mobile telephony and in 2012 it earned about 500 million euro through licensing those patents to industry players. Thanks to its R&D investment and its active participation in standards for wireless, Nokia kept well ahead of the wireless technology curve for over a decade. While some of its technical breakthroughs never made it into mass production, the company was first to market with many widely adopted mobile device technologies. Between 1995 and 2007, it was clearly the pacesetter that other mobile companies tried to emulate.

Ironically Nokia's market-leading mobile phone manufacturing processes coupled with the control over mobile device adoption features by wireless carriers combined to reduce the company's incentives for disruptive innovation.

Nokia achieved its success by focusing on core competencies in manufacturing processes and device technology. But as the market came to value software and applications as much as hardware platforms, it lost momentum. Its alignment with wireless carriers put Nokia out of touch with other important players in the market, reducing its ability to create partner ecosystem and limiting Nokia's flexibility in business model innovation through app stores and data services.

New entrants like Apple seized the opportunity to create an end-to-end business that linked its device to application developers and direct to the consumer through an Apple-controlled App Store. A more innovative ecosystem and business model strategy freed Apple from relying solely on iPhone hardware sales to generate revenue and gave it a new source of bargaining power in making deals with wireless carriers.

In September 2013, almost exactly 3 years after the arrival of Stephen Elop as the CEO expected to lead Nokia's turnaround, Elop's former employer Microsoft announced that it was acquiring Nokia's mobile devices business for $7.9 billion. It was the end of the race for a former mobile pacesetter.

References

Best, J.: 'Android before Android': The long, strange history of Symbian and why it matters for Nokia's future. ZDNet (April 4, 2013). URL: www.zdnet.com/android-before-android-the-long-strange-history-of-symbian-and-why-it-matters-for-nokias-future-7000012456/

Editor: Nokia chief strategist: "Apple will remain a niche player in smartphones, just like in computers". Mac Daily News Blog Posting (November 30, 2009). URL: macdailynews.com/2009/11/30/nokia_chief_strategist_apple_will_remain_a_niche_player_in_smartphones/#AFVEt6L6hz3rQ2Du.99

Elop, S.: Stephen Elops 'burning platform' memo. Wall Street Journal (February 9, 2011). URL: blogs.wsj.com/tech-europe/2011/02/09/full-text-nokia-ceo-stephen-elops-burning-platform-memo/

Nokia: About Nokia. Nokia On-Line (2013). URL: www.nokia.com

Nokia: Conversations by Nokia. Nokia Blog Posting (September 10, 2010). URL: conversations.nokia.com/2010/09/10/stephen-elop-to-join-nokia-as-president-and-ceo/

Ocock, T.: Symbian OS – one of the most successful failures in tech history. TechCrunch (November 8, 2010). URL: techcrunch.com/2010/11/08/guest-post-symbian-os-one-of-the-most-successful-failures-in-tech-history-2/

Reinhardt, A.: Nokia's magnificent mobile-phone manufacturing machine. Business Week On-Line (August 2, 2006). URL: www.businessweek.com/stories/2006-08-02/nokias-magnificent-mobile-phone-manufacturing-machine

Vanjoki, A.: The fight back starts now. Conversations by Nokia, Nokia Blog Posting (July 2, 2010). URL: conversations.nokia.com/2010/07/02/the-fightback-starts-now/

Chapter 5
Amazon Fast Tracks Transformation

As a company, we are culturally pioneers, and we like to disrupt even our own business. Other companies have different cultures and sometimes don't like to do that. Our job is to bring those industries along.
  Jeff Bezos, Amazon Founder and CEO (Levy 2011)

Amazon opened its web site to the public in 1995, at a time when buying a product on the Internet was far from simple. Few of the product search options, customer reviews and user interfaces that today's ecommerce shoppers take for granted were fully developed. Each online store made its own idiosyncratic design choices about how to illustrate its products, where to locate the shopping cart icon, and how many steps were needed to complete a purchase. Often such decisions were based on the aesthetic of the site designer rather than on any data about keeping visitors engaged or the best way to manage the checkout process so that customers didn't abandon their online shopping carts. As Jeff Bezos put it in his 1997 letter to Amazon shareholders, "This is Day 1 for the Internet, and, if we execute well, for Amazon.com" (Bezos 1997).

Amazon emerged as a pacesetter for online retail innovation with a combination of data driven best practices and partnering strategies to pinpoint and solve the critical challenges of Internet shopping. Consumers expected ecommerce transactions to be convenient and trustworthy. But few of the early web stores delivered. In comparison to its competitors, Amazon made it easy to proceed from simply browsing, to selecting a product, to completing a purchase. Rather than relying on arbitrary design decisions, Amazon implemented a process of prototyping multiple web site options and relentlessly testing the performance of each option in the process of converting visitors into buyers.

In September 1997 Amazon introduced its 1-Click® buying option. Once customers registered their preferred payment information and delivery address with Amazon, they could complete subsequent purchases immediately after selecting a product without having to enter more details or go through any added checkout procedure.

1-Click® shopping meets the criteria that Peter Drucker applied innovation—simplicity. "To be effective an innovation has to be simple, and it has to be focused....

M.J. Cronin, *Top Down Innovation*, SpringerBriefs in Business,
DOI 10.1007/978-3-319-03901-5_5, © The Author 2014

Indeed the greatest praise and innovation can receive is for people to say 'This is obvious! Why didn't I think of it it's so simple!' Even the innovation that creates new users and new markets should be directed to a specific, clear, and carefully designed application" (Drucker 2002).

1-Click is focused on achieving one simple goal—getting the Amazon visitor to make a purchase. Today it seems obvious that streamlining the buying process would encourage more online shoppers to complete their transactions. Indeed Amazon's patent for the underlying business processes that enabled its 1-Click button was bitterly challenged by Barnes & Noble and other online competitors who claimed that a one step online buying process was too basic to merit patent protection. The Amazon 1-Click patent withstood these challenges giving Amazon a significant pacesetter advantage. With a patent in hand, Amazon could have denied every other retailer the opportunity to offer such a streamlined check out. Instead, it decided to leverage its invention by licensing the 1-Click process to potential competitors such as Toys"R"Us and Target. The decision to license was an early indicator that Amazon's long-term ambitions reached beyond online retailing to encompass a wide variety of business models and industry ecosystems.

The 1-Click process was just one component of Amazon's strategy for defining and implementing online retail best practices. Innovative solutions for boosting brand recognition, winning consumer trust, and motivating purchases included pioneering a world-class online Affiliates program, hosting detailed customer reviews for each product, and developing a product recommendation engine. As each new component of the online store was developed, alternative web page elements were prototyped and tested live, generating a continual flow of data to determine what worked best. The company's relentless focus on using data in its decision making was embodied in all of Amazon's internal processes.

The first part this case will analyze how Amazon's key online retail innovations established it as the industry pacesetter and set the stage for its long term ecommerce market leadership. In recent years, Bezos' founding vision of creating "the biggest store on earth" has expanded to include an appetite to transform industries as diverse as publishing and media, logistics and IT services. Part 2 will focus on how Amazon extended its innovation impact, making a nimble transition from an online retail pacesetter strategy to become a Transformative Architect for multiple industries.

Setting the Pace for Online Retail Best Practices

Making online retail simple is not easy. Ensuring that the Amazon web site delivered a great customer experience required constant measurement, experimentation and readjustment behind the scenes. Amazon employees implemented a regimen of A/B testing to track customer reaction to different ways of organizing the home page, displaying product information and other design options. This tracking and testing provided Amazon with vital data to fine tune every aspect of the online shopping experience in ways that produced higher conversion rates and more revenue per customer.

As an early ecommerce store, Amazon had an opportunity to set expectations for online buyers, establishing best practices in an emerging industry sector. Many of the company's early design decisions set the standard for customer experience across the online retail store sector. At the same time, Amazon faced formidable challenges in building trust in its new brand. With no physical store location to reassure customers, web visitors considering a purchase needed convincing that the process was safe and reliable.

Amazon's early innovation focused on converting online visitors into customers through the optimization on its website design, navigation and shopping cart experience. Given the high costs of maintaining a huge selection of products, it had to succeed at stimulating regular purchases and creating customer loyalty to the Amazon brand. This was a tall order. In his first letter to shareholders after Amazon's 1997 IPO, Bezos acknowledged that ecommerce was still in its infancy. To track overall corporate success and shareholder value Amazon's focus would be on measuring "customer and revenue growth, the degree to which our customers continue to purchase from us on a regular basis, and the strength of our brand" (Bezos 1997).

What Jeff Bezos would in later years refer to as the company's "culture of metrics" was also in evidence in Amazon's early use of the website analytics data to determine which of the many web site options proposed by staff and management were actually paying off for the company. An anecdote from Greg Linden, a software engineer who started working at Amazon in 1997, demonstrates how experimentation and data trumped opinion—even when the opinion was held by top managers.

According to Greg, he found time to do a bit of unofficial prototyping of different options for the Amazon checkout page. In particular,

> I loved the idea of making recommendations based on the items in your Amazon shopping cart. Add a couple things, see what pops up. Add a couple more, see what changes. The idea of recommending items at checkout is nothing new. Grocery stories put candy and other impulse buys in the checkout lanes. Hardware stores put small tools and gadgets near the register. But here we had an opportunity to personalize impulse buys. It is as if the rack near the checkout lane peered into your grocery cart and magically rearranged the candy based on what you are buying.

Greg built a quick prototype, launched it on an internal test site and asked colleagues for feedback. Senior marketing managers protested, concerned that making new offers during checkout might distract shoppers and increase the rate of shopping cart abandonment—a key metric for retail success. Greg was told to stop experimenting, but he took some personal risk and put his recommendation prototype into the regular Amazon testing sequence to get some real data on how shoppers responded. The test results demonstrated that personalized recommendations didn't increase cart abandonment—instead they led to additional purchases, boosting sales revenues significantly. Based on this data, Greg's idea became a priority for implementation and a standard feature for Amazon (Linden 2006).

This anecdote crystallizes how data-driven decision making enabled Amazon to set the pace for early online retail innovation. A process that allowed employees to test their ideas and get results data in real time encouraged innovation from every level of staff. Experimental web site changes could be put into production if they led to desired results and quickly eliminated if they did not. This process opened the door to thousands of small improvements that kept Amazon ahead of its online competition and reduced the risk of experimentation. The constant comparison of alternative options, and opportunities to optimize based on data analysis, encouraged thinking about broader improvements and innovative programs.

Amazon continued to be a data-driven and experiment based company as it harvested the insights from tracking online customer interactions. But website optimization was just one part of retail success. Amazon needed to apply the same rigor to its logistics and fulfillment practices.

Mastering Operational Challenges

By 2000 Amazon had achieved many of the online customer experience and growth goals outlined by Jeff Bezos in the company's 1997 annual report. It had weathered the dot.com crash to become the market leader in online retail, with over $2.76 billion in sales and more than 20 million active customers in the US and internationally. Amazon had established its brand as a trusted source for an expanding line of products that included electronics and household goods as well as books, music and film.

However, this rapid growth was not sufficient to make Amazon a profitable business. In 2000 the company posted a loss of $1.4 billion, a level of red ink that threatened Amazon's ability to keep expanding and fed skepticism about its prospects for long-term survival.

Amazon had become the pacesetter on the web, but it had not mastered the traditional retailer challenges of inventory management, logistics and product fulfillment. It was time for the online optimizer to subject its logistics and back end processes to a comprehensive data-driven make over. Accordingly, Jeff Bezos announced a new focus for the company in 2001—"the march to profitability."

To kick start the process, Amazon developed computer models of all its internal operations. For the first time, it put systems in place to track the cost and profit margin (or loss) incurred for every product it sold, collecting myriad details on packing and shipment costs, the frequency of damages in shipment, product returns and 47 other factors that influenced profit margins. The resulting torrent of data forced staff to rethink many of the company's fulfillment assumptions.

In May 2001 Amazon agreed to allow Saul Hansell, a New York Times reporter, to spend 3 days at company headquarters in Seattle, observing Amazon's internal staff discussions about implementing a new set of metrics that focused on process and operations efficiency. According to Hansell's report, he arrived at a time when the data analysis had uncovered widespread logistics problems that were blocking Amazon's quest for profitability, "Executives found that more than 10 percent of the products sold from the electronic, kitchen and tool departments lost money, while 5 percent of the book, music and video products were losers" (Hansell 2001).

After much internal debate, Amazon decided to stop stocking certain unprofitable products. In future, the company would use distributors for direct shipping of less frequently ordered items including book titles and electronics. It would expand its third party merchant ecosystem by encouraging outside sellers to handle items with low profit margins.

The drive to rationalize product inventory and implement operations algorithms to make warehousing, logistics and fulfillment more efficient cut company losses dramatically. Amazon achieved its goal of profitability in 2002. It continued the push to rationalize all aspects of its inventory, logistics and fulfillment management, using its data to constantly improve the decision process. By 2005, Amazon had become a pacesetter in online fulfillment based on data analysis. As described by Bezos in 2005,

> Many of the important decisions we make at Amazon.com can be made with data. There is a right answer or a wrong answer, a better answer or a worse answer, and math tells us which is which. These are our favorite kinds of decisions.
>
> Opening a new fulfillment center is an example. We use history from our existing fulfillment network to estimate seasonal peaks and to model alternatives for new capacity. We look at anticipated product mix, including product dimensions and weight, to decide how much space we need and whether we need a facility for smaller "sortable" items or for larger items that usually ship alone. To shorten delivery times and reduce outbound transportation costs, we analyze prospective locations based on proximity to customers, transportation hubs, and existing facilities. Quantitative analysis improves the customer's experience and our cost structure. (Bezos 2005)

In turning its logistics weakness into a strong point, Amazon created a foundation for innovation in its fulfillment processes. It also built an infrastructure that could be shared with third parties to generate additional revenue and increase profit margins. The strategic use of internal capabilities to architect new business opportunities was to become a hallmark of Amazon's Transformative Architect strategy in the coming decade.

Boosting the Impact with Ecosystem Partners

In 2000, Amazon opened its online store platform, shopping cart and order processing capabilities to third-party merchants with the launch of its Marketplace service. Bezos recognized that there was some risk inherent in welcoming potential competitors into Amazon's web site to access its growing customer base. But there were also strategic and business benefits. The Amazon Marketplace charged commission fees for each order that it processed for its new merchant partners, providing a much-needed boost to revenue and profit margins. Although there was a certain amount of competition and overlap of items for sale, for the most part merchants who joined the Amazon ecosystem were selling the less frequently ordered or low margin goods that Amazon could not handle profitably. This expanded the scope of Amazon retail coverage, bolstering the company's leadership position as the world's largest online store while continuing to streamline its logistics costs. As more merchants joined the Marketplace, Amazon also benefitted from having more customer interactions and purchase decisions to analyze. Access to the data generated by each merchant sale provided expanded visibility into the popularity of millions of products, giving Amazon a broader view of customer preferences and purchasing patterns.

Opening the Amazon platform to third party users created a major new ecosystem of business partners. There are now over two million third party merchants selling tens of millions of products on Amazon. Analysts estimate that as of 2013 these sellers account for more than 40 % of all Amazon web store sales.

The 2005 launch of Amazon Prime as a customer loyalty program featuring free 2-day shipping on all Amazon direct orders created an ecosystem for leading customers. A 2010 *BusinessWeek* article describes Amazon Prime as "the most ingenious and effective customer loyalty program all in all of ecommerce if not retail in general" (Stone 2010).

Beyond the boost in revenues and sales that comes from Prime members frequent shopping, Amazon can look to its Prime customer ecosystem to be early adopters for its new products and services. For example, Amazon bundles a free 6-month Prime membership with the Kindle Fire. Prime members get free access to Amazon content from books to films and TV series, whether accessed on the Fire or when streaming to a TV or another mobile platform using an Amazon app. That's a significant value, especially as Amazon's media content libraries are comparable to those of Netflix and Hulu which require monthly subscription fees to access. New buyers of the Kindle Fire get an automatic Prime membership to jumpstart their use of the device with free content as well as motivation to shop for all the Amazon physical products available with free shipping to Prime members.

Triggering Industry Transformation

Amazon emerged from the collapse of dot.com bubble as a well established brand and world leading online retailer. It had transformed itself internally into an operations logistics and fulfillment best practice leader, improving its own margins on every product sold and enabling a business model through its Amazon Marketplace.

Jeff Bezos, however, was not content to limit his company's innovations to the online retail sphere. He saw multiple opportunities to leverage the capabilities that Amazon had mastered in its first decade to provide even more ambitious and disruptive services. Amazon had architected its web infrastructure to provide a sharable and scalable ecommerce platform that it could leverage to create new business opportunities. This strategy became a model for Amazon to enter new industries as a transformational infrastructure and platform provider in its second decade.

In his 2011 interview with Wired Magazine, Bezos explained the decision to launch Amazon Web Services (AWS) in 2006 as a natural outgrowth of Amazon's own needs for scalable Internet services,

The problem was obvious. We didn't have that infrastructure. So we started building it for our own internal use. Then we realized, "Whoa, everybody who wants to build web-scale applications is going to need this." We figured with a little bit of extra work we could make it available to everybody. We're going to make it anyway—let's sell it. (Levy 2011)

It was far from obvious to investors in 2006 that Amazon could generate profits by offering web services. Analysts were becoming impatient with Amazon's heavy technology investments and mixed financial results. The concept of selling low-priced slices of computing time over the internet to individual clients and small customers seemed several steps away from Amazon's core retail capabilities. In the short run, the AWS launch meant that Amazon was once again putting profitability on the back burner in favor of new growth opportunities, investing its capital in building up its infrastructure at the same time that it was aggressively expanding its dedicated physical warehouse locations around the world.

A November 2006 cover story in *BusinessWeek* characterized AWS as "Jeff Bezos' Risky Bet" describing Amazon's new strategy as follows,

Bezos wants Amazon to run your business, at least the messy technical and logistical parts of it, using those same technologies and operations that power his $10 billion online store. In the process, Bezos aims to transform Amazon into a kind of 21st century digital utility. It's as if Wal-Mart Stores Inc. had decided to turn itself inside out, offering its industry-leading supply chain and logistics systems to any and all outsiders, even rival retailers. (Hof 2006)

One can almost hear Bezos chiming in to remind readers that it would probably be 5 or more years before his AWS and warehouse infrastructure bets started to pay off. Since 2010 Amazon has spent over $13 billion on 50 new warehouses around the world, bringing its global total to 89 as of 2012. The launch of AWS was just the start of Amazon's determination to dominate the emerging computer infrastructure service sector now known as cloud computing. By 2013, cloud-based services were among the fastest growing areas of Information Technology spending worldwide and were generating an estimated $3 billion in annual revenue for Amazon. Gartner Research estimates that Amazon has five times the computing capacity as its 14 closest competitors in cloud services, an impressive measure of the company's lead in this high-growth industry sector (Edwards 2013).

Extending the Transformation of Logistics and Fulfillment

Amazon Fulfillment, also launched in 2006, had none of the digital innovation appeal of cloud computing. Providing warehouse space and shipping services to smaller companies seemed the opposite of innovative. Weren't companies trying to shed their expensive physical infrastructure? As Bezos saw it, so many companies wanting to outsource inventory and fulfillment meant that the sector was ripe for innovation and transformation. This vision put Amazon 5 years ahead of the wave in seeing the transformational synergy between big data, cloud infrastructure and automating logistics management.

Amazon heavy investment in both computing and logistics infrastructure makes sense in the context of its joint goals—to set the pace for online retail practices and to transform multiple industry sectors. The massive infrastructure investment redefines online retail best practices for speed and cost of delivery and raises the bar for competitors to Amazon's diverse lines of business. It motivates companies to sign up for AWS and Fulfillment in order to benefit from Amazon's advanced infrastructure; in turn the fees paid by these customers help to support the cost of that infrastructure. Finally, the participation of millions of smaller merchant partners in the Amazon ecosphere extends Amazon's transformational reach.

Raising the bar has pressured major retail players to change their own processes. Wal-Mart is still the $469 billion gorilla of the physical retail world, compared to Amazon's total sales of $61 billion in 2013. But Wal-Mart conceded that it had failed to keep pace with best practices in online fulfillment. Wal-Mart online shipment costs are running from $5 to $7 per parcel, compared to Amazon averages of $3–4. Speed of deliveries are lagging as well, motivating Wal-Mart to spend over $430 million tailoring its warehouses and logistics systems to ecommerce requirements (Banjo 2013).

With its 2012 purchase of Kiva Systems, a supply chain robot maker, Amazon took the next step in transforming logistics and fulfillment. Forbes calls Kiva "an integral part of the migration to Cloud Computing" noting,

> It's old news that the Internet unleashed a new form of commerce in cyberspace. The next evolutionary step for the Internet is the Cloud that connects the cyber world directly to the physical world. Here Amazon is on the leading edge precisely because they have from inception successfully straddled these two domains. (Mills 2012)

From Disrupting Publishers to Transforming Publishing and Media

If analysts were skeptical about Amazon's foray into web services and warehouse fulfillment services, they were even harsher in predicting that Jeff Bezos was courting disaster with the 2007 launch of the Amazon Kindle e-reader. Few expected that Amazon, a novice in hardware development, would be able to match the high bar that Apple had set for the innovative design of consumer devices.

Unfavorable comparisons between the Kindle and the elegant design of Apple's iPod and iPhone were inevitable, but these comparisons misinterpreted the Kindle's role in Amazon's broader plans for re-architecting the publishing industry. As an online retail pacesetter with an early focus on book sales, Amazon was well acquainted with the vulnerabilities of traditional publishing models. Its insistence on discounting book prices reduced the publishers' profit margins but this was offset in part by its rapidly growing customer base which made it the single largest distribution channel for most publishers. The most serious disruption occurred at the bookstore level, threatening the survival of small local shops and large chains like Borders and Barnes & Noble alike. As local stores shut their doors, Amazon's dominance as a sales channel reduced the publisher's bargaining power over digital book prices and set the stage for an industry-wide transformation. The Kindle was just one part of Amazon's plan to architect an end-to-end platform for digital publishing and content distribution.

Amazon applied much the same strategy in launching the Kindle and its ebooks content library that it had implemented so successfully with the Amazon.com store debut in 1995. It focused on the key value propositions of selection, low price and convenience to entice customers to adopt the device. Like Amazon.com, the Kindle store launched with a huge selection of low priced best seller titles based on Amazon agreements with publishers to price their ebooks at $9.99 each. Kindle buyers could also download hundreds of thousands of free classic titles. Original works by self-published authors were free or priced from $0.99 to a few dollars per title. This mix of best sellers, free classics and low-priced original works created a long tail of content and customer preference data that leveraged the recommendations systems Amazon had already perfected for its online retail customers. Kindle users had access Amazon's collection of millions of customer reviews along with person-alized recommendations for what to read next on the Kindle. To make the buying decision as easy and as tempting as possible, Amazon also let Kindle users sample ebook chapters for free.

Amazon made its Kindle reading application available for all leading smartphones and tablets as well as Kindle devices. Since all Kindle ebook purchases are stored on the Amazon cloud, Amazon has the advantage of controlling and tracking how that content is used at the same time as making it convenient for consumers to access their Kindle book purchases on any device. To provide a seamless reading experience, the Kindle app even remembers the place you stopped reading on one device and picks up in the same spot when you resume reading on another device.

The Kindle was optimized to connect users instantly and seamlessly to Amazon's digital content ecosystem. Book purchase was streamlined to reflect the 1-Click shopping experience. Once a Kindle was registered, it wirelessly connected to the Amazon cloud allowing Amazon users to immediately link to their existing Amazon accounts and payment credentials—and start buying ebooks. Kindle devices did not need to be hardware design leaders. They delivered a premier reading experience with innovative E-Ink screens to reduce eyestrain and reader-friendly features such as search, annotation, word lookup, highlighting, and note taking along with an extended battery life of weeks rather than hours of use per charge.

Readers responded positively. By 2010, sales of Kindle ebooks had overtaken hardcover print books on Amazon and by the following year, Kindle ebooks were outselling all types of printed books on Amazon. Amazon was well underway in establishing its dominance in digital publishing, a rapidly growing market that was estimated to generate over $10 billion in revenue in the US alone by 2016.

The Kindle became the foundation for a multi-device family that now includes multiple versions of the Kindle reader and the Fire tablet. Bezos calls these devices an integrated media service. With Amazon-connected devices, consumers are always just one click (or one screen touch) away from browsing, buying, reading, watching or otherwise interacting with some Amazon product, content or service.

Amazon is working to expand those opportunities for interaction, building mobile and social services into all of its customer touch points. Like Netflix, it is also investing in creating original entertainment content, hiring award-winning writers and producers to produce original films and TV-type series. With the integration of digital production, distribution and consumption, Amazon has created an end-to-end service aimed at disrupting the media industry with the strategy for transformation that has been so effective in retail and in publishing.

Conclusion

In his wide-ranging interview with Wired in 2011, Bezos summed up his personal innovation style as follows, "I like invention. For me, it feels like the rate of change on the Internet today is even greater than it was in 1995. It's hard for me to imagine a more exciting arena in which to invent" (Levy 2011).

Jeff Bezos has infused Amazon with a drive to innovate, transform, and dominate markets. Under his leadership, the biggest bookstore on earth has come a long way since 1995. He clearly plans to keep on growing and transforming the industries in Amazon's widening sphere. As of October 2013, Amazon's corporate mission statement encompasses the entire ecommerce universe: *We seek to be Earth's most customer-centric company for four primary customer sets: consumers, sellers, enterprises, and content creators* (Amazon 2013).

Amazon has integrated the core building blocks of innovation in different combinations to drive the company's long-term goals. It built in data analytics, A/B testing and experimentation at every stage of its online retail growth in order to determine which online store innovations paid off and which ones should be abandoned. It developed ecommerce best practices as a pioneer of online retail, inventing foundational web store components to make online shopping easier and more convenient.

When its profitability was on the line, the company embraced a major rethinking of its retail operations on the back end, including inventory management logistics warehousing and supplier partnerships for direct shipping. It established an innovative, scalable platform that was architected to support selling Amazon logistics and fulfillment services to third-party merchants.

Amazon Web Services and Cloud Computing infrastructure design reflect the same integration of technology, process and service innovation to generate new market-leading businesses. Today Amazon is a force to be reckoned with far beyond online retail. Its dominance in cloud computing and logistics and its transformation of digital publishing and media have made it a respected and often feared architect of industry transformation.

This transformation strategy continues to unfold, as Amazon is now well positioned to tackle a number of interconnected industry sectors including online payment, app development, IT services, media content, TV and film production. Amazon Studios, which Bezos describes as a completely new way of making movies, uses crowdsourcing to put a social spin on its media production decisions. Amazon invites consumers to watch the pilot programs of potential new series on its web site and then vote for the pilots that Amazon should produce as new series. It is also inviting consumers to submit their ideas, scripts and short videos for consideration. These innovations contrast to the more traditional media production model that Netflix has pursued with its House of Cards and other original productions. It's not clear if Amazon's more disruptive strategy will win favor with a mass market audience. But Bezos continues to experiment with opportunities for transforming yet another industry.

With over 150 million visitors each month, a cadre of active Amazon Prime customers, and millions of ecosystem partners for its digital content and merchant services platforms, any Amazon innovation that extends into adjacent market sectors will have significant revenue and disruptive potential for the impacted industries. The Amazon Media Group, launched in 2012, grew out of a long-standing practice of hosting third party ads on Amazon product pages. In its first year as a stand-alone Amazon service, the Media Group proved itself as a fast-growing competitor to advertising agencies and digital ad networks, with potential to generate over $1 billion in revenue for Amazon by 2015.

Amazon's transformative strategy reached into payment services with the 2013 launch of "Login and Pay with Amazon" a new online checkout service that rivals PayPal's market leading online checkout service. PayPal is processing an estimated $100 billion worth of on online payments each year for its merchant clients, so Amazon's entrance into the online payment market has huge growth potential.

What next? There are rumors of a mobile phone with unique 3D features in the wings. Whether or not Amazon ever produces the next big online entertainment blockbuster, or releases an innovative mobile phone, it is clear that the market awaits its next announcement with anticipation that there will be disruption ahead.

For Jeff Bezos, it is still day 1 for transformation opportunities.

References

Amazon: Corporate mission statement. Amazon On-Line (2013). URL: www.amazon.com

Banjo, S.: Wal-Mart's E-Stumble. Wall Street Journal B1 (June 19, 2013)

Bezos, J.: Letter to shareholders. Amazon On-Line (1997). URL: www.amazon.com

Bezos, J.: Letter to shareholders. Amazon On-Line (2005). URL: www.amazon.com

Drucker, P.F.: The discipline of innovation. Harvard Business Review 80(6), 95–103 (2002)

Edwards, J.: Amazon has reached a staggering level of dominance when it comes to cloud computing. Business Insider (August 27, 2013). URL: www.businessinsider.com/amazons-aws-market-share-and-revenues-2013-8-ixzz2hLhZ7tYa

Hansell, S.: Listen up! it's time for a profit; a front-row seat as Amazon gets serious. New York Times On-Line (May 20, 2001). URL: www.nytimes.com/2001/05/20/business/listen-up-it-s-time-for-a-profit-a-front-row-seat-as-amazon-gets-serious.html?pagewanted=all&src=pm

Hof, R.D.: Jeff Bezos' risky bet. BusinessWeek On-Line (November 12, 2006). URL: www.businessweek.com/stories/2006-11-12/jeff-bezos-risky-bet

Levy, S.: Jeff Bezos owns the web in more ways than you think. Wired On-Line (November 13, 2011). URL: www.wired.com/magazine/2011/11/ffbezos/all/1

Linden, G.: Early Amazon: Shopping cart recommendations. Geeking with Greg Blog Post (April 25, 2006). URL: glinden.blogspot.com/2006/04/early-amazon-shopping-cart.html

Mills, M.P.: Amazon's Kiva robot acquisition is bullish for both Amazon and American jobs. Forbes On-Line (March 23, 2012). URL: www.forbes.com/sites/markpmills/2012/03/23/amazons-kiva-robot-acquisition-is-bullish-for-both-amazon-and-american-jobs/

Stone, B.: Whats in Amazons box? Instant gratification. BusinessWeek On-Line (November 24, 2010). URL: www.businessweek.com/magazine/content/1049/b4206039292096.htm

Chapter 6
Stanford Spins Out a Higher Education Tsunami

> We're not the people that develop the next version of the iPhone or whatever, we're the people who develop the product that nobody thought of, the concept, the discovery that nobody has made yet....We try to keep the innovation flame going.
> John Hennessy, President of Stanford University (Lashinsky 2012)

"Develop the product that nobody thought of"—that sounds like a clarion call for radical innovation. Is it also an appropriate mission for a non-profit school that was founded in the nineteenth century? When that school is Stanford University, John Hennessy's emphasis on the business of innovation makes perfect sense. Stanford shares an elite academic ranking with its east coast Ivy League peers, but its geographic location next to Silicon Valley has imbued the campus—professors, students, and the top administration—with an entrepreneurial and innovative zeal that is as pronounced as it is unusual in higher education. In keeping with founder Leland Stanford's exhortation to prepare students for personal success and direct usefulness in life, Stanford describes itself as a wellspring of innovation. Its web site proudly notes that in the last 70 years over 39,000 businesses owe their start to the inventions and entrepreneurial aspirations of its faculty, administrators and students.

John Hennessy, who became Stanford's tenth president in 2000, is unabashed about encouraging the entrepreneurial focus of the university that he joined in 1977 as an electrical engineering faculty member. Hennessy has supported scores of major technology-based innovations for industry during his time at Stanford. In some cases he has contributed to them directly. Professor Hennessy's innovations in the development of RISC computer architecture flowed from the academic research lab into his entrepreneurial success as a cofounder of MIPS Computer Systems, a microprocessor design company, and back into his position as a professor. Unlike many universities, where direct faculty involvement in off-campus business ventures is tightly regulated if tolerated at all, Stanford goes out of its way to facilitate faculty entrepreneurial efforts. As Stanford's top executive, Hennessey has fostered the close relationship between Silicon Valley and University faculty and researchers, serving on the board of Google and other companies founded by Stanford alums. The boundaries between the university and industry remain flexible as many of

M.J. Cronin, *Top Down Innovation*, SpringerBriefs in Business,
DOI 10.1007/978-3-319-03901-5_6, © The Author 2014

Stanford's leading researchers follow in Hennessey's own footsteps by founding and running companies for a period of time without completely severing their ties to Stanford or giving up their faculty appointments.

The university's close partnerships with industry and its track record of spinning out highly successful business ventures reach back to the days of Hewlett-Packard, Fairchild, Cypress Semiconductor and Silicon Graphics. Enterprises with Stanford roots include Cisco, Intuit, Sun Microsystems and dozens of pharmaceutical and biotech companies. In the ecommerce, social, and mobile era Stanford is more active than ever as a hotbed of transformational ideas, an incubator and launch pad for tech-based business models. The list of startups turned online powerhouses under the leadership of Stanford faculty and alumni reads like a Who's Who of today's digital disruptors featuring eBay, Google, LinkedIn, Yahoo, Zillow and Netflix. Inventions that have been patented by Stanford researchers and faculty include foundational breakthroughs in fields that vary from digital music synthesizing and the Google page rank algorithm to recombinant DNA and high-speed data transmission over phone lines (Stanford 2013).

Stanford comes in for its share of criticism for its pro-entrepreneurial encouragement of the pursuit of business success. A New Yorker article in 2012 reflected on the school's reputation as "Get Rich U" and speculated about possible conflicts of interest that stem from the two way industry and investor connections that bring entrepreneurs and venture capitalists into the classroom and encourage university faculty and students to be active participants in startups and well established Silicon Valley enterprises (Auletta 2012). Stanford's Silicon Valley setting and close connections with some of the world's most successful entrepreneurs may well blur the line between academia and industry. Arguably, however, this blurring also makes Stanford a particularly apt testing ground for experiments in teaching and learning that are designed to shake up long-held assumptions about the role of the traditional classroom and campus settings.

This case analyzes Stanford's role as a recent architect of transformation in the higher education sector. It focuses on the emergence of the Massive Open Online Course (MOOC) as a potentially disruptive educational model. Stanford opened its traditional on-campus courses as a testing ground for the faculty who designed and taught the early MOOCs, including Daphne Koller and Andrew Ng who went on to cofound Coursera and Sebastian Thrun, founder of Udacity. These two MOOC companies have come to epitomize the potential of massive open online courses for transforming higher education as well as the challenges faced by colleges and universities when adopting such platforms.

The Making of a MOOC

Even before John Hennessy's arrival on campus, Stanford University was experimenting with technology to extend selected programs and course content reach. In 1969 the school broadcast its engineering graduate courses to off-campus students using satellite television. In 1996 Stanford Online became the first university Internet system that incorporated text, graphics, audio and video for course access. The university was an early participant in Apple's iTunesU, uploading podcasts of Stanford guest speaker presentations and faculty class lectures to iTunes. In 2008 Stanford's Engineering Everywhere became one of the first free sites to offer complete video-based courses and materials on demand, including ten free computer science and electrical engineering courses.

During the past decade, Stanford was part of a large cohort of colleges and universities that made their course content available for free through iTunes, academic consortia, or school websites. The Massachusetts Institute of Technology pioneered the OpenCourseWare (OCW) initiative in 2002, to share MIT course materials such as syllabi, lecture notes and assignments. Free OCW content has expanded to include materials from over 2,100 MIT courses, attracting over a million web visits every month. The MIT and Stanford programs fit into an even larger context of thousands of earlier online course offerings from all types of US colleges and universities. Distance learning has a long history in the US with significant participation by both for-profit and non-profit institutions.

According to the 2013 Babson report on online learning in US higher education, the majority (71.7 %) of US colleges and universities reported offering some of their courses online as far back as 2002. At that point online learning was already an important part of the US educational landscape, serving approximately 1.6 million students. In the past 10 years, the number of students taking at least one higher education course online has increased to 6.7 million representing an all time high of 32 % of US students in 2012 (Allen and Seaman 2013).

What makes today's MOOCs different from these prior online course programs? Is that difference enough to justify the widespread predictions that massive open online courses will transform higher education? A closer look at the launch of massive online courses at Stanford helps to illustrate what distinguishes such courses and how Stanford's experimental and entrepreneurial atmosphere helped to shape today's leading MOOC platforms.

In some respects, the 2011 offer of free online access to an elective course in artificial intelligence taught by Stanford's Computer Science Department faculty was part of a long-standing Stanford practice of sharing course content with the public. The particular course offering did not initially get any encouragement from Hennessy, however, because the faculty members behind it didn't ask for university permission in advance. The administration got involved only after more than 10,000 people had registered. Opening up their classroom to thousands of online students with no payment and no questions asked could be a career-ending move for faculty on many college campuses. At Stanford, it caused just a minor administrative perturbation as the worldwide impact of this particular faculty experiment became clear.

Perhaps if Sebastian Thrun and Peter Norvig, the teachers of the Introduction to Artificial Intelligence course, had been full time Stanford faculty members rather than Google executives teaching as adjunct professors, they would have sought permission before launching a free online enrollment site for their course and planning for streaming video lectures and interactive tests. Characteristically for Stanford, both instructors held full time industry positions in Silicon Valley and were teaching the computer science elective as adjunct professors. Norvig was director of research at Google, and Thrun, who had previously been a full time tenured professor and director of Stanford's Artificial Intelligence Laboratory, had left campus to take a position with Google X, a research division of Google focused on experimental projects such as developing self-driving vehicles.

Given Stanford's long-standing encouragement of faculty entrepreneurship, however, their adjunct status may have not had much to do with their decision to proceed without bothering to seek any official university approvals. Thrun had been inspired by hearing Salman Khan, founder of the non-profit Khan Academy, describe the impact that Khan Academy science and math video modules are having on students in elementary and high schools. According to Khan the self paced videos have encouraged teachers to adopt a "flipped classroom" style of learning in which students learned course basics by watching the videos on their own, freeing up in-class time for discussion and help with areas that needed more explanation. This learning option benefitted advanced students as well as those who needed extra time to review the videos at home and get individual help in class. In his presentation, Khan noted that the free online video modules were providing access to a fast-growing global community of learners.

According to an article in *Wired*, after hearing the Khan speech, Thrun was inspired to revisit his ideas for an interactive teaching/learning platform combining streaming video lectures, interactive testing, and other cloud-based learning activities to support college level courses online. The course he and Norvig were currently teaching, "Introduction to Artificial Intelligence" seemed like a good test case. It had enrolled 177 Stanford students for the traditional on-campus version the previous year, so was already designed for a lecture plus assignment format. The free online course would cover the same material with the same level of assignments and tests, just assembled as a collection of streaming videos with interactive quizzes and grading. Thrun arranged for the video lectures and quizzes to be hosted on the Amazon cloud and linked them up to a learning management infrastructure from a startup he was already working with (Leckart 2012).

Since Stanford wasn't being asked to host the online version of the course, it didn't seem necessary to request any official campus support. Students completing the online version would not get any Stanford or other college credit, just a certificate of accomplishment for finishing the course and a report on how their performance ranked compared to others in the class. Norvig estimated that there might be a maximum of 2,000 who would register and far fewer who would actually finish the course.

Registration for Artificial Intelligence started slowly in summer of 2011 with some low key announcements to publicize the course. As word of the open registration opportunity to take a Stanford course spread around the web, thousands more registrations poured in. Soon there were tens of thousands of students signed up. Inevitably, Stanford administration took note that the Artificial Intelligence faculty had created an alternative course platform. As Thrun recalls, it took a series of meetings and compromises to address the university administration and legal department concerns. But none of the administrators raised any fundamental objection to the concept of a free and open online version of his course. The main sticking points were that Stanford wanted clarity that the course would not offer Stanford academic credit or any kind of formal certification. Since Thrun had already announced that there would be a certificate for those who finished with a passing grade, the wording was renegotiated for it to become a statement of achievement rather than a certificate.

The Artificial Intelligence course launched online with 160,000 students registered—an enrollment that far exceeded anyone's expectations and that threatened to overwhelm the online streaming and interactive infrastructure that Thrun's start up partner had designed for a maximum of 10,000 users. Using the social network component of the course, students worked out ways to solve some of the issues that the massive enrollment created. They answered each others' questions, posted detailed examples and even designed open web sites to practice the programming exercises that hadn't been fully integrated with the course materials. Before the end of that first semester, Thrun had decided that his embryonic start up idea deserved full time attention. By year end the new company has raised its first round of venture funding, hired a team of engineers, and been re-named Udacity.

Daphne Koller and Andrew Ng, full time Stanford Computer Science professors, made the same leap from campus to startup after teaching their own classes on an open MOOC platform in fall 2011. Ng's course in Machine Learning attracted over 100,000 students online. In a pattern that has characterized most subsequent MOOCs, the majority of students dropped out before completing the course. In Machine Learning, more than half never completed the first homework assignment and only 13,000 made it to the finish line. Regardless of how many online students had mastered the fundamentals of Machine Learning, Ng and Koller were convinced that MOOCs were a bridge to the future of global education. At the beginning of 2012, they became the cofounders of Coursera, flush with $16 million in funding from prestigious Silicon Valley venture capital firms, and buttressed with commitments from Stanford and other elite university partners to offer free courses through the Coursera MOOC platform.

Stanford's support and association gave academic credibility to the MOOC model and helped to create an extraordinarily fast start for both companies. The founders clearly benefitted from their Stanford faculty status in attracting media coverage, venture capital funding, and the willingness of other leading universities to partner with their fledgling educational ventures. Even Stanford's reputation would not be enough, however, to sustain the rapid growth of Coursera and Udacity if the MOOC model did not represent significant potential for transformational improvements in teaching and learning. Both companies have leveraged their technology innovations to develop faculty-student interactions, course content and social media platforms that presage fundamental changes for higher education.

Catalyzing Open Online Capabilities

"Taking the Measure of MOOCs," an October 2013 report in the *Wall Street Journal*, crystallizes promising innovations of the open course platforms, along with some major MOOC challenges, in a few telling numbers. On average, a MOOC student who posts a question on the course discussion forum will get a response from fellow students within 11 min. The median amount of time a student typically spends watching a 6–9 min video in a sample of science or math MOOC courses is 6.29 min. The pass rate for a traditional on-campus Introduction to Engineering course at San Jose State was 55 %; when a MOOC component was added to the same course the pass rate increased to 91 %. Other comparisons of traditional course and MOOC outcomes at San Jose State tell a different story. Only 50 % of the students in an all-online MOOC course on elementary statistics in spring 2013 earned a C or better grade, compared to 76 % of the students in the classroom-only course. Most striking, a single MOOC course can collect 230 million data points about the online activities of the participating students (Fowler 2013).

What do these and other metrics tell us about the effectiveness and future potential of MOOC-based learning? On the plus side, a number of early MOOC metrics have confirmed that the tight integration of student discussion forums, social media and collaborative incentives support an interactive learning experience on the MOOC platforms that extends beyond the content presented through faculty videos. Discussion forums encourage active participation by the students. Getting a quick answer to a course question helps students to keep up with the material and reinforces their sense of being part of a community. The data shows that there are also benefits for the helpful students who post answers to questions—the most active social media participants are the most likely to complete the course and to do well in assignments and tests.

While most student interactions take place online, the discussion forums provide a variety of options for course participants to arrange for local meetings, to set up face-to-face as well as virtual study groups, and to contribute their personal study guides and related materials to the forum. The level of energy that some students put into their courses and the amount of help that they provide to fellow students in responding to questions and posting supplementary material provide important added value and become a self reinforcing aspect of success in the course. Students who get quick and helpful responses to their questions feel more committed to making an effort to finish the course. Students who provide answers to questions are more heavily invested in the overall course outcome and statistically they are significantly more likely to succeed in all the course requirements.

MOOC platforms are leveraging interactivity and collaboration in more formal ways to collect and utilize student-to-student comments on class assignments for peer grading of projects, further redefining the roles of teacher and student in online learning. These efforts have experienced the growing pains of early stage technology, but they show the promise of scaling to support human feedback on assignments and papers for otherwise unmanageable size groups. Real time automated grading of objective questions and multiple choice quizzes are already built into the platform. MOOCs are also developing algorithms and artificial intelligence concepts to expand the capabilities of computer grading to handle qualitative essays and assignments. Even with the limitations of multiple choice quiz and objective test formats, the instant feedback provided when students take a test, and the encouragement to go back to review materials to improve the test score, is a powerful motivator.

The San Jose State administration announced a pause to reassess its partnership with Udacity after seeing disappointing results from the spring 2013 semester, only to find that there was a huge rebound in the success rate in the same set of MOOC courses during the summer session. In the summer 2013 MOOC course in elementary statistics, 83 % of the students earned a grade of C or above, compared to the baseline of 76 % of students taking the classroom version. Some of the variability was due to a different mix of students enrolled in each semester. Udacity also made some changes in the course implementation to improve interaction and provide more student support. These early outcome data show the difficulty of making any evaluation about the long-term value of MOOCs at the current stage of experimentation of most campuses. It also shows that the variables that influence outcomes are still not well understood and can be tweaked in relatively small ways to dramatically impact the outcome for better or worse.

Whether or not San Jose State expands its MOOC offerings and achieves even better results over time, the availability of detailed outcome data and the use of this data for benchmarking to constantly measure and improve student learning provide a powerful advance over current learning assessment practices. MOOCs can measure thousands of details regarding each student's level of involvement, interaction and mastery of the subject from day to day. This in itself is a breakthrough which helps to establish a new standard of accountability.

Data Drives the Future

The data points that are being collected in each of the courses and the focus on data-driven learning improvement are among the most promising aspects of MOOC courses and a significant advance over the traditional classroom and prior online education programs. MOOC platforms are designed to monitor and analyze individual student engagement and measure student understanding of course concepts in ways that even the most attentive classroom teacher cannot hope to match. Faculty members are limited to a very small number of observations of student learning during a typical in-class lecture and discussion. These observations are a subjective complement to the handful of traditional tools for evaluating student performance in an on campus course—grading the homework, course assignments, and exams. The final course grade is a rough measure, at best, for ranking student performance and quantifying learning outcomes.

In a MOOC course format, every online action can be tracked, including student participation in the discussion forums, how they interact with the videos, when and how long they view course content, as well as the specifics of the scores they get on quizzes and exams and assignments. Analysis of all of this data provides new insights into the learning process and how to improve it; insights that can be applied to improving future course design.

The data about such a massive group of online learners tackling the same problems at the same time also provides unique direct feedback to the teacher about any lack of clarity in teaching content, presentation, or in the test questions. If thousands of students get a certain question wrong, professors and the course platform designers can zero in on what might have caused the high error rate. Was it something specific about the wording of that particular question, or does it point to the need to cover the relevant material more thoroughly in future lectures and assignments?

Data on peer interactions and peer grading provide ideas for future course innovations. Each course provides millions of indicators about what works best for improving outcomes for particular types of students and content. Analyzing and feeding this insight back into the course design process allows participating schools and faculty members to continually measure learning outcomes, eventually making adjustments in real time depending on the actions of enrolled students at different stages in a specific course.

In the future, smarter algorithms and automated data mining tools can be used to scan across a massive archive of courses and student responses, develop predictive models of where learning problems might arise or where extra resources and examples are needed, and proactively create personalized online instruction by inserting the most appropriate materials based on the learning styles of individual students. Coursera and Udacity are nowhere near achieving that goal today. But the data that has been collected and shared with participating institutions is a key factor in continual improvement for the course delivery platform and for the interactions among all students, and with the professor, to harness the power of predictive data to an even greater degree.

As is typical of disruptive new entrants, many aspects of today's MOOC courses are inferior to the traditional classroom experience. MOOC founders understand that today's offerings are far from perfect. They envision a positive slope of improvements that leverages the data generated through massive student participation to redesign the course platform over time, creating a personalized and dynamic, interactive experience for each individual student.

All the data being collected are not yet being to put work to analyze how the course content interacts in real time with student learning behavior, certainly not at a level comparable to the constant A/B testing and website optimization of ecommerce leaders like Amazon. The MOOC companies are too new and are growing too quickly to have achieved data mining and predictive analytics best practices across all aspects of their platform.

As future versions of the MOOC platforms combine insights from massive course data points with technology innovations, the MOOC executives expect that the courses they offer in a few years will have measurably superior learning outcomes. At some point in the not too distant future, if MOOC platforms and course content design continue to receive focused attention and investment, they will match or even exceed the traditional classroom lecture experience for many college courses. Even at this early stage of improvement, the MOOC startups have unique advantages to attract customers—in the case of Coursera and Udacity, access to free courses designed for self-paced learning that are taught by professors from the world's leading universities.

Other priorities are demanding the founders' attention in the short term, including the expansion of courses and building the ecosystem partnerships that will lead to revenues and new business models for both Coursera and Udacity. The two companies are taking their MOOC platforms in two different directions that reflect the founder's strategies for creating value, and provisioning course content.

Coursera and Udacity founders have repeatedly stated that the first priority is creating the infrastructure and ecosystems that will support educational transformation rather than implementing a predetermined revenue model. They are following a digital startup strategy of scaling to serve millions of users to prove their value proposition before worrying about demonstrating long term profitability. As the MOOC platforms continue to improve and the education partner ecosystems expand, Coursera and Udacity will be better positioned to determine what aspects of their innovation are likely to generate revenue.

Coursera has focused on building its ecosystem of academic course offerings by pursuing partnerships with hundreds of educational institutions around the world. Coursera relies on this ecosystem of elite higher education partners to provide course content for its MOOC platform and leaves many of the specifics of course design to the academic partner. As a result, Coursera course offerings have been criticized as uneven, with some universities simply retrofitting old videos of traditional classroom lectures with a few embedded quizzes. One early course provided by Georgia Tech had to be withdrawn from Coursera because of technology glitches in coordinating assignments and quizzes with the lectures (Kolowich 2013).

Udacity takes a more hands-on approach to content, partnering directly with professors to produce courses that are designed to take full advantage of their MOOC platform. This gives the company more control over the design and interactivity of each course and allows Udacity to customize and tweak its courses more easily when improvements are indicated, as with the San Jose State example. The disadvantage is that its model of direct involvement limits the number and the types of courses that Udacity makes available. Producing high-quality online content and integrating interactive learning features into an online course is very expensive and continues to be a work in progress for Udacity which has only 28 courses available for fall 2013 registration. In comparison, Coursera offered 482 courses provided by 91 educational partners as of October 2013.

Coursera and Udacity are not the only new entrants competing to transform higher education through MOOC technology. The non-profit EdX consortium was launched by MIT and Harvard University in 2011 with its own MOOC platform. The EdX platform hosts courses taught by Harvard and MIT professors plus courses from a handful of EdX partner universities. It offers many of the same hosting features, course development tools, analytics and social student interfaces as Coursera and Udacity. In September 2013, Google and EdX announced a partnership to work together on further development of Open EdX. The partners plan to make the open source MOOC platform available through a new site, MOOC.org, in 2014. They expressed hope that the availability of an open platform paired with app interfaces and development tools will encourage third party developers to build innovative new features for Open EdX, creating an applications and developer ecosystem that will add value for institutions and teachers who use Open EdX to design a course and present content to online students.

Udacity has embarked on a similar ecosystem strategy, enlisting Google as a partner along with AT&T and NVIDIA, in its recently announced Open Education Alliance. From the point of view of transforming classroom learning, the burgeoning interest in MOOC partnerships and the accompanying investment by major tech industry players in MOOC technology reflect the impact that MOOC innovations are making on college level teaching and learning quite aside from the long term commercial success of Coursera and Udacity as business ventures.

The jury is still out on how appropriate and effective MOOCs will be for certain types of subjects and learning environments, including credit-bearing college courses and graduate degree programs. There is no question, however, that MOOCs are already impacting the higher education sector. MOOC courses will continue to reach millions of students around the world, and MOOC platforms will improve and change over time to make access to college level courses more accessible to a global market.

Combine these factors with a widespread sense of crisis in higher education, and the likelihood of industry disruption leading to transformation is high. The next section examines the confluence of forces facing higher education that make it ripe for change along with the factors that make MOOC models likely to catalyze a transformation of higher education during the next decade.

Higher Education: Ripe for Transformation

At the beginning of 2013, Moody's Investors Services downgraded the outlook for the entire US higher education sector from stable to negative. Moody's annual industry outlook emphasized that for the first time it included even the top-ranked research universities and elite private colleges in its unfavorable ranking, citing mounting financial pressure on all major university revenue sources. Among the economic pressures on higher education, Moody's cited continued price sensitivity to tuition costs that limited tuition increases, a dip in the number of US high school graduates, reduced public university funding and overall federal funding, and increases in student loan defaults. "The US higher education sector has hit a critical junction in the evolution of its business model," noted a Moody's analyst and "most universities will have to lower their cost structure" to remain financially viable (Moody's Investors Services 2013).

Vocal government and public criticism of college tuition cost increases, combined with over $1 trillion in outstanding student loans in the US, have fueled a growing skepticism among students and their families about the value of a traditional college education relative to the potential earnings of students after graduation. Smaller colleges are already confronting declining application pools and enrollment rates. Top higher education administrators are saying openly that traditional higher education in the US cannot survive without major transformation. Either this transformation will come from within or it will be forced on colleges through government and state controls, regulation and funding cuts.

Digital learning innovation offers an opportunity to break through longstanding cost and growth constraints. The emergence of MOOC courses that could benefit students and improve learning outcomes while radically reducing the cost of traditional course delivery and degree programs has come at the perfect juncture to attract serious consideration by higher education decision makers. The majority of colleges and universities, however, are still far from making a major commitment to MOOC implementation. Media coverage of open online courses and other online learning innovations highlight the early adopters, but only 2.6 % of the 2,800 US colleges and universities in the 2012 Babson survey were offering a MOOC. While chief academic officers at the majority of institutions report that online learning is critical to their long-term strategy, only 27.8 % of academic leaders responded that MOOCs are a sustainable method for offering courses. Students and faculty are the biggest skeptics. According to the Babson Survey, only 30.2 % of faculty members express acceptance of the value and legitimacy of online education. At schools with no online offerings, faculty acceptance sinks to 10 %.

University administrators and faculty alike are tracking Georgia Tech's pioneering decision to offer its well-respected Master's degree in Computer Science using Udacity's MOOC platform in partnership with AT&T. The online Master's degree will be equivalent to Georgia Tech's long established Computer Science graduate program except for a radical cost reduction. Instead of paying around $44,000 for the traditional on-campus degree, MOOC students will pay only $6,000 to complete their coursework online. Georgia Tech reports an influx of qualified applications from all 50 states in the US and from 80 other countries for the launch of the program which will offer its first MOOC courses in January 2014 (Belkin 2013).

The use of MOOC innovations to offer more cost-effective degree opportunities from well-established universities will inevitably pressure the cost model at institutions which do not consider or adopt the efficiencies available from interactive online learning platforms. Whether or not the faculty skepticism about MOOCS is reduced or simply ignored, many university administrators are seeing the value and the necessity for this type of disruptive change on their own campuses.

Even the well-endowed and entrepreneurially inclined Stanford University is calculating the cost benefits of replacing traditional (and increasingly unpopular) lecture format classes with the interactive forums, online lectures and self-correcting quizzes that characterize MOOC offerings. John Hennessy sees "bending the cost curve" as a critical advantage of MOOC courses and supports continued experimentation with online learning on his own campus as well as for the industry sector as a whole. Why would Stanford as one of the leading US universities want to facilitate the disruption of the higher education establishment in the US? Hennessy summarized his recommended MOOC strategy in a keynote presentation at the Computing Research Association in 2012 as follows, "My thought: better to face the future than to hide from it. Be the disrupter; not the disrupted" (Hennessy 2012).

References

Allen, E., Seaman, J.: Changing course: Ten years of tracking online education in the United States. Babson Survey Research Group (January, 2013)

Auletta, K.: Get Rich U: There are no walls between Stanford and Silicon Valley. should there be? New Yorker On-Line (April 30, 2012). URL: www.newyorker.com/reporting/2012/04/30/120430fa_fact_auletta

Belkin, D.: First-of-its-kind online masters draws wave of applicants. Wall Street Journal p. A.7 (2013).

Fowler, G.: An early report card on MOOCs. Wall Street Journal p. R.1 (2013)

Hennessy, J.L.: The coming tsunami in educational technology. Keynote Speech, The Computing Research Association, 40th Anniversary Conference at Snowbird (July 22, 2012)

Kolowich, S.: Georgia tech and Coursera try to recover from MOOC stumble. Chronicle of Higher Education (February 4, 2013). URL: chronicle.com/blogs/wiredcampus/georgia -tech-and-coursera-try-to-recover-from-mooc-stumble

Lashinsky, A.: Will the world's greatest startup machine ever stall? Fortune Magazine On-Line (June 20, 2012). URL: tech.fortune.cnn.com/2012/06/20/stanford/

Leckart, S.: The Stanford education experiment could change higher learning forever. Wired On-Line (March 20, 2012). URL: www.wired.com/wiredscience/2012/03/ff_aiclass/all/

Services, M.I.: Press release. Moody's On-Line (January 16, 2013). URL: Higher-Education-sector-changed-PR_263866

Stanford: Wellspring of innovation. Stanford On-Line (2013). URL: www.stanford.edu/group/wellspring/

Chapter 7
Conclusion

The pace of technology change is only going to accelerate. Disruptive innovation, propelled by cloud computing, big data, ecommerce, ubiquitous mobile and social connectivity, and the rising generation of born-digital entrepreneurs, is the new normal for managers in every industry sector. For an increasing number of organizations, growth prospects and survival will depend on their innovation strategy. Executives can decide to lead with innovation, or respond to it, but they can no longer ignore it.

The Top Down Innovation Model highlights the central role of top level managers in the success or failure of innovation within their organizations. Executives must articulate a clear and consistent innovation vision, identify the critical metrics that will be used to measure progress toward that vision, implement the infrastructure required for integration and analysis of those critical performance areas, then empower and reward employees for making data-driven innovations that lead to measurable improvements on those areas. These steps may sound obvious, but they are far from the norm for innovation management.

The foundation of Top Down Innovation management is analyzing key data points to measure, benchmark, and continually improve performance and return on innovation across the organization. But according to the 2013 global executive study and research project by MIT Sloan School of Management, the majority of global enterprise executives have not even defined the key performance indicators that should be used to track the outcomes of their current investment in digital innovation. Only 26 % of the companies surveyed had set up clear metrics to define innovation success (Fitzgerald et al. 2013). It is unlikely that the 74 % of organizations that have no structures in place to measure performance are getting a strong return on their innovation investments.

Both external and internal factors are at play in an executive's decision about which of the three core innovation strategies is the best organizational match. Managers should regularly scan their industry and competitive landscape to assess disruptive trends and identify the competitors that are implementing Transformative Architect and Nimble Pacesetter strategies. If there is already a dominant Transformative Architect, it may be productive to align with that company by participating in their ecosystem or adopting a compatible technology infrastructure. If there is a successful Nimble Pacesetter, evaluate the technology bet they have made. Is that technology peaking and likely to be replaced by an emerging alternative over the next several years? This could provide an opening for a competitive Pacesetter strategy. If the competitive landscape has no dominant innovators, the Power Practitioner approach of adopting digital innovations from adjacent industries will provide competitive differentiation.

Internal factors include organizational culture, profitability and available resources, technology capabilities, and executive leadership style. Is the organization reliant on legacy technology and proprietary systems or is it already using cloud computing and open platform infrastructure? Have data sources and analytics tools been integrated to provide management and employee access to all key data and performance metrics? Does the company have resources available to devote to new digital initiatives?

If the chief executive and the shareholders of well-established companies have very different appetites for risk, or if the company is struggling to maintain profitability or lagging in comparison to industry best practices it may be premature to select the more risky Pacesetter and Architect strategies. If top management sees a compelling opportunity for disruptive new products and services, the path to market leadership still starts with achieving best practice performance across the organization.

The top level characteristics for executives implementing a Power Practitioner, Nimble Pacesetter, or a Transformative Architect strategy are summarized in Fig. 7.1.

Power Practitioner organizations are aligned around data-driven decision making on a daily basis. Managers and employees focus on meeting and exceeding best practice milestones. The organization recognizes and rewards focused innovation that results in measurable performance improvements. This consistent and disciplined approach to innovation across the organization results in profitable long term growth.

Executives who prefer the higher risk strategies of Nimble Pacesetter and Transformative Architect must ensure that their organizations have internalized the Power Practitioner culture of benchmarking, best practices and performance improvement. Without high-performing, industry leading processes, a company risks losing its market lead to more focused competitors.

Nimble Pacesetters are confident enough to bet on emerging technology trends that will support innovative business models, products and services. If the technology succeeds, Pacesetter executives will invest in capturing market share while optimizing their processes to keep up with market demands. Pacesetters should build up barriers to new entrants with patents for key innovations, personalized customer services and strong brand differentiation. The major challenge for the confident managers of successful Pacesetter companies is transitioning away from their preferred technology once it has reached its peak. If, like Nokia, they delay making this transition they risk falling victim to the Innovators Dilemma.

Transformative Architect strategies require a visionary leader with an appetite for risk. Executives who pursue this strategy must be able to win over shareholders, employees and potential ecosystem partners to their vision for the future of the industry.

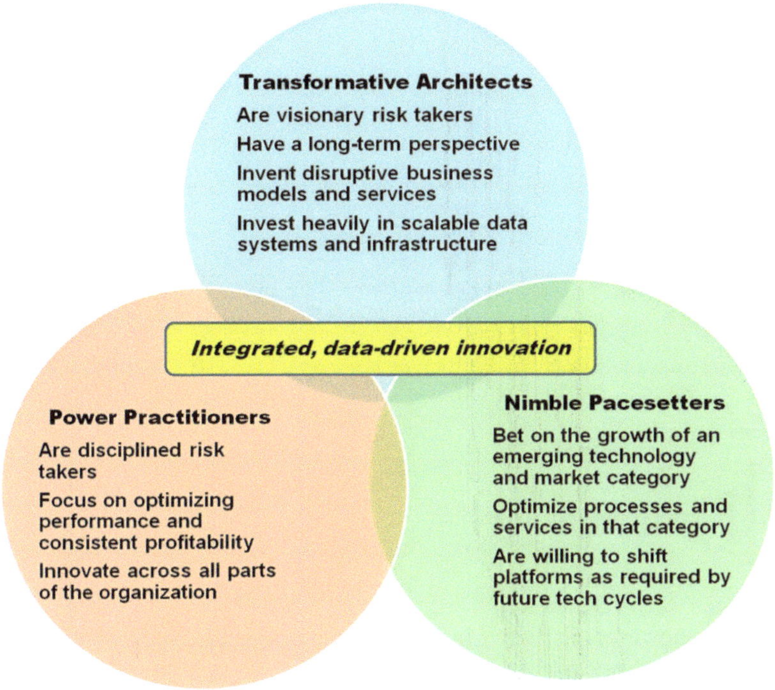

Transformative Architects

Are visionary risk takers

Have a long-term perspective

Invent disruptive business models and services

Invest heavily in scalable data systems and infrastructure

Integrated, data-driven innovation

Power Practitioners

Are disciplined risk takers

Focus on optimizing performance and consistent profitability

Innovate across all parts of the organization

Nimble Pacesetters

Bet on the growth of an emerging technology and market category

Optimize processes and services in that category

Are willing to shift platforms as required by future tech cycles

Fig. 7.1 Requirements for core innovation strategies

Their high-risk ambitions must be matched with a zeal for optimizing performance as well as designing the required technical infrastructure for scalable ecosystems As we saw with Bill Ford and Ford Motor, disruptive vision without a Power Practitioner foundation can lead to business disaster. Ford had to bring in a strong practitioner CEO to turn the company around to restore profitability before it could pursue more radical innovation opportunities.

Transitioning Among Innovation Strategies

Once a strategy is selected and articulated, top managers must stay focused and align all components of the integrated innovation framework around the key metrics for success. Selecting and committing to a core strategy clarifies the role of innovation within the organization and sets the stage for achieving measurable goals. Nimble Pacesetters need to focus on their chosen market sector. They won't thrive if top management switches course to pursue every disruptive technology and new category opportunity that presents itself. Power Practitioners can be most effective in a relatively stable industry sector with a large global market. They are not likely to succeed in an attempt at transformation of adjacent industries.

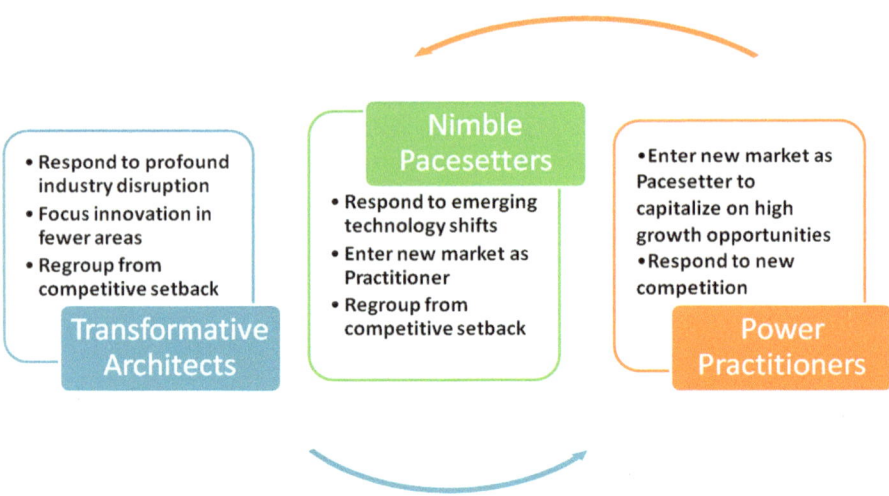

Fig. 7.2 Paths for core strategy transition

At times, however, disruptive new entrants, major industry and technology shifts, or rapidly changing market opportunities may provide compelling reasons for executives to shift from one core strategy to another. A change in the company's executive leadership is another occasion for reevaluating the core innovation strategy. The new top executive may deliberately steer the organization toward another strategy that is a better match for his or her vision for the future.

Steve Jobs' return to Apple is a notable example of an executive-led strategic pivot toward a more transformative core strategy. IBM's decision to move from hardware to services and eventually to sell off its PC division is a move in the other direction, from Transformative Architect of the mainframe and PC eras to a Power Practitioner technology services provider. Reed Hastings decision to move away from DVD rentals to focus Netflix on the streaming video sector meant transitioning from being the Pacesetter market leader in the DVD category to being a Power Practitioner in the highly competitive streaming video, Internet TV market.

Figure 7.2 illustrates potential paths of transition between core strategies along with some of the major reasons that organizations shift from one innovation strategy to another.

Before considering any strategic transition, it is important to assess the organization's current innovation practices. The following questions will help executives to check the pulse of innovation and pinpoint areas that need management attention to measure up to best practices.

1. *Can every employee and stakeholder describe your organization's core innovation strategy and their role in making it a success?*
   If the core innovation strategy has never been discussed within the organization, this will be an obvious no. Even if there is a mission statement that addresses

innovation, top managers should regularly engage employees in dialogue about innovation at the individual and department level.

2. *What are the three most important metrics that your company uses to benchmark the success and impact of innovative projects?*
   In the majority of organizations, these metrics have not been identified and consistently measured. Making this a priority will jump start the management discussion of innovation strategy and put the company ahead of its competition.

3. *What performance measures, competitive benchmarks and other data points do top executives, managers and employees review daily?*
   Executives who insist on a data centric, integrated approach to performance measurement across all parts of the organization are setting the stage for innovation to provide measurable benefits.

4. *What role do ecosystem partners, suppliers, and your customers have in discussing and contributing to decisions about your company's products and services?*
   Extending the innovation discussion to customers and partners will provide new data as well as new ideas and incentives for improvement.

5. *How are employees rewarded for suggesting and implementing innovative improvements?*
   Align the reward system around data driven decisions and measurable outcomes.

In the shifting technology and competitive landscape that every manager faces, there is no single silver bullet to ensure innovation success. Putting data at the heart of decision making, selecting and articulating a core innovation strategy, and implementing an integrated framework are best practices that will increase the return on innovation activities.

Reference

Fitzgerald, M., Kruschwitz, N., Bonnet, D., Welch, M.: Embracing digital technology. MIT Sloan Management Research Review (2013)